Java网络编程进阶

从BIO到RPC

肖 川 编著

清华大学出版社
北 京

内容简介

本书用于学习Java网络通信的设计和开发，系统地介绍了Java网络通信的多种技术，由浅入深地阐述了多种通信技术的基础知识、主要模型以及实际可用的案例，使读者能有效地掌握Java网络编程的相关知识，并具备以Java编程来处理和解决网络通信问题的能力。

全书共8章和1个附录。第1～4章依次介绍BIO、NIO、AIO及Netty编程技术，每章均提供3个案例的设计和编码；第5章与读者分享RESTful应用轻量级框架Jersey的使用经验；第6章介绍Web服务消息推送规范SSE，并基于Jersey的SSE机制实现订阅-发布功能以及一个可重入的分布式锁；第7章自行设计了一个RPC框架并进行代码实现；第8章开发了两个简单常见的应用；附录解答了Java开发时常见的若干问题。

本书可作为高等院校"网络程序设计"课程的教材，也可供相关领域的技术人员参考。

图书在版编目（CIP）数据

Java网络编程进阶：从BIO到RPC/肖川编著. —北京：清华大学出版社，2021.2 (2024.8重印)
ISBN 978-7-302-57575-7

Ⅰ. ①J…　Ⅱ. ①肖…　Ⅲ. ①JAVA语言－程序设计　Ⅳ. ①TP312.8

中国版本图书馆CIP数据核字(2021)第028857号

责任编辑：刘向威　常晓敏
封面设计：文　静
责任校对：胡伟民
责任印制：刘　菲

出版发行：清华大学出版社
网　　址：https://www.tup.com.cn, https://www.wqxuetang.com
地　　址：北京清华大学学研大厦A座　　**邮　　编**：100084
社 总 机：010-83470000　　**邮　　购**：010-62786544
投稿与读者服务：010-62776969，c-service@tup.tsinghua.edu.cn
质量反馈：010-62772015，zhiliang@tup.tsinghua.edu.cn
课件下载：https://www.tup.com.cn，010-83470236
印 装 者：三河市君旺印务有限公司
经　　销：全国新华书店
开　　本：185mm×260mm　　**印　　张**：13.5　　**字　　数**：332千字
版　　次：2021年4月第1版　　**印　　次**：2024年8月第3次印刷
印　　数：3001～3600
定　　价：59.00元

产品编号：090783-01

前言

随着网络及Java技术的发展，分布式和微服务成了企业信息部门在技术选型时的首选。网络编程技术是分布式系统开发的基石，无论是使用现有的微服务框架开发业务应用，还是自行研发底层的服务框架，了解、掌握底层的网络编程技术如AIO、NIO等，对开发者来说都是必不可少且多多益善的技术修炼。基础不牢，地动山摇，丰富自己的技术知识栈将为开发者的职业生涯提供更加广阔的发展空间。

本书特色

网络编程是一门实用型技术，必须理论和实践相结合。本书在阐述理论知识或设计思路时，辅以更为直观的图解，使其更易理解；本书亦用大量的篇幅展示落地实用的Java代码并对其进行分析和解释；通过案例的开发和分析，本书还向读者展示了Java网络技术与Java其他技术如线程、同步器、泛型、反射等的关联使用。

读者对象

- Java程序员
- 分布式系统架构师
- 高校“网络程序设计”课程的学生
- 其他对Java网络编程感兴趣的读者

本书内容

基于学以致用的原则，本书通过8章内容和1个附录来介绍Java网络编程相关的技术。

第1～4章依次介绍基于BIO、NIO、AIO及Netty的编程技术，每种技术实现3个案例的设计和编码。

第5章与读者分享RESTful应用轻量级框架Jersey的使用经验，包括同步请求及应答、异步请求及应答、基本认证和授权以及如何替换某些部件，这些经验使得Jersey应用的开发更加高效和鲁棒。

第6章介绍Web服务消息推送规范SSE的使用，基于Jersey的SSE机制实现订阅-发布功能以及一个可重入的分布式锁。

第7章自行设计一个RPC框架并进行代码实现，阐述了设计方案，并对关键代码进行解释。

第8章开发两个简单常见的应用，一个是基于WebSocket的聊天室；另一个是邮件发送程序。

附录解答了Java开发时常见的若干问题。

本书所关注的网络编程技术符合业界当今主流，可以有效提高读者的Java网络编程技术水平及核心竞争力。

源码和课件

本书的Java源码和PPT课件可从https://github.com/Adairiver/jnp下载，亦可在清华大学出版社的官方网站上获取。

致谢

感谢我的家人。本书的写作占用了大量的业余时间，没有家人的支持和理解，这本书不可能完成。

感谢清华大学出版社的编辑刘向威博士。因为向威博士的一直鼓励和帮助，本书才会如此顺利地出版。

由于笔者的水平有限，书中难免有不足之处，还望读者海涵和指正。非常期待能够得到广大读者的反馈，在技术之路上互勉共进。

肖　川

2020年9月

目录

第 1 章　BIO 1

1.1　Socket 通信模型 1

1.2　完善通信框架 3

1.3　升级 write 与 read 6

1.4　案例 1：传输字符串的会话 9

1.5　案例 2：传输对象的会话 11

1.6　案例 3：传输文件的会话 15

习题 20

第 2 章　NIO 21

2.1　NIO 模型 21

2.2　NIO 服务端框架代码 22

2.3　NIO 客户端框架代码 25

2.4　ByteBuffer 及其在 NIO 中使用的问题 26

2.5　NIO 的分帧处理 29

2.6　案例 1：传输字符串的会话 32

2.7　案例 2：传输对象的会话 35

2.8　案例 3：传输文件的会话 40

2.9　设计多线程服务器 48

习题 56

第 3 章　AIO 57

3.1　异步操作概述 57

3.2　AIO 服务端框架代码 59

3.3　AIO 客户端框架代码 60

3.4　AIO 的分帧问题 61

3.5 案例 1：传输字符串的会话 64
3.6 案例 2：传输对象的会话 67
3.7 案例 3：传输文件的会话 70
习题 76

第 4 章 Netty 77
4.1 Netty 的使用模型 77
4.2 Netty 的入站与出站 79
4.3 服务端框架代码 81
4.4 客户端框架代码 82
4.5 ByteBuf、分帧以及 ChannelHandler 链 83
4.6 案例 1：传输字符串的会话 85
4.7 案例 2：传输对象的会话 92
4.8 案例 3：传输文件的会话 97
习题 107

第 5 章 Jersey 108
5.1 概述 108
5.2 案例 1：对象资源的操作 110
5.2.1 服务端基本框架 110
5.2.2 客户端基本框架 112
5.2.3 逐项添加 URI 功能 114
5.3 案例 2：异步请求与异步应答 123
5.3.1 服务端基本框架 123
5.3.2 客户端基本框架 124
5.3.3 逐项添加 URI 功能 124
5.4 案例 3：基本认证和授权 139
5.4.1 服务端基本框架 139
5.4.2 客户端基本框架 140
5.4.3 服务端认证项 141
5.4.4 客户端认证项 145

5.5 案例 4：替换某些部件 146
5.5.1 替换 JSON 解析器 147
5.5.2 替换 Servlet 容器 148
5.5.3 替换 Web 服务器 149
5.5.4 完全剥离 Spring 151
习题 153

第 6 章 SSE 155

6.1 SSE 概述 155
6.2 订阅-发布功能 156
6.2.1 服务端代码 156
6.2.2 客户端代码 160
6.3 实现分布式锁 164
6.3.1 分布式锁服务端 166
6.3.2 分布式锁客户端 169
6.3.3 分布式锁的使用 173
习题 174

第 7 章 实现 RPC 框架 175

7.1 RPC 框架概述 175
7.2 框架的客户端设计 177
7.2.1 序列化器 177
7.2.2 代理层 179
7.2.3 通信层 181
7.3 框架的服务端设计 184
7.3.1 序列化器 184
7.3.2 反射层 186
7.3.3 通信层 188
7.4 服务消费者 191
7.5 服务发布者 193
习题 194

第8章　两个简单应用 **195**

8.1　WebSocket应用 195

8.2　邮件发送程序 200

习题 203

附录 **204**

第 1 章 BIO

BIO(blocking I/O)称为阻塞式 I/O,可以用在请求并发量不是很高的网络环境。理解 BIO 是掌握 Java 网络通信的起点,而且便于对比以后的非阻塞式 I/O。本章对应的源代码为附件中的 bio 项目。

1.1 Socket 通信模型

图 1-1 所示的 Socket 通信模型是一个基本模型,与阻塞、非阻塞无关,与同步、异步也无关。

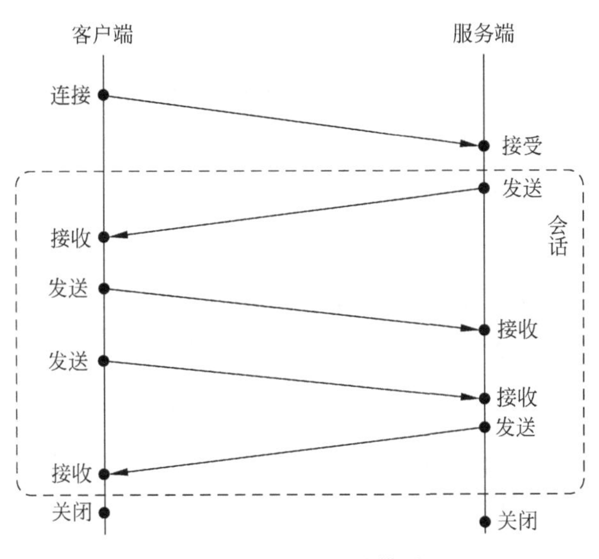

图 1-1 Socket 通信模型

双方通信的流程包括 3 个步骤。

(1) 建立连接。服务端开启服务端口,等待客户端的连接,客户端发起连接请求,服务端接受请求,双方建立连接。

(2) 进行会话。建立连接后,客户端与服务端之间传输业务数据。客户端发送,同时服务端接收;或者服务端发送,同时客户端接收。会话期间双方可以进行多次发送与接收。

(3) 关闭连接。会话完毕时,客户端和服务端各自关闭连接。

根据上述通信流程,编写 BIO 网络通信的雏形代码,包括两个类,Server0 类代表服务端,Client0 类代表客户端。

代码 1-1:类 Server0

```
1  public class Server0 {
2      private int port;
3      public static void main(String[ ] args) {
4          Server0 server = new Server0(8001);
5          server.service();
6      }
7      public Server0(int port){
8          this.port = port;
9      }
10     public void service(){
11         try {
12             ServerSocket serverSocket = new ServerSocket(port);
13             System.out.println("服务器开启服务端口" + port + ",等待连接请求……");
14             while(true) {
15                 Socket socket = serverSocket.accept();
16                 System.out.println("接受客户端的连接,开始会话……");
17                 session(socket);
18                 socket.close();
19                 System.out.println("会话结束,服务端关闭连接。");
20             }
21         }catch (Exception e){
22             e.printStackTrace();
23         }
24     }
25     public void session(Socket socket) throws Exception{
26         System.out.println("与客户端进行会话,接收或者发送数据……");
27     }
28 }
```

代码 1-1 的第 15 行是运行的阻塞点,当没有来自客户端的连接请求时,accept()方法将等待,直到有新的连接请求来到。

代码 1-2:类 Client0

```
1  public class Client0 {
2      private String serverIP;
3      private int port;
4      public static void main(String[ ] args) {
5          Client0 client = new Client0("127.0.0.1", 8001);
```

```
6            client.communicate();
7        }
8        protected Client0(String ip, int port){
9            this.serverIP = ip;
10           this.port = port;
11       }
12       public void communicate() {
13           InetSocketAddress isAddr = new InetSocketAddress(serverIP, port);
14           Socket socket = new Socket();
15           try {
16               socket.connect(isAddr);
17               System.out.println("客户端连接成功,开始会话……");
18               session(socket);
19               socket.close();
20               System.out.println("会话结束,客户端关闭连接。");
21           }catch (Exception e){
22               e.printStackTrace();
23           }
24       }
25       public void session(Socket socket) throws Exception{
26           System.out.println("与服务端进行会话,接收或者发送数据……");
27       }
28 }
```

代码 1-2 的第 16 行是运行的阻塞点，客户端会一直试着连接直至连接成功或者连接超时。

这个通信雏形过于简化，例如，服务器采用单线程，也没有考虑超时设置，代表会话的 session(.)函数具体实现有待完善。

1.2 完善通信框架

代码 1-1 和代码 1-2 的基本框架需要在超时控制及多线程方面进行完善。首先增加超时设置，有 3 种超时，即连接超时、读超时、写超时。超时设置可以有效地控制阻塞时间。

1. 连接超时的设置

方法 socket.connect(SocketAddress endpoint, int timeout)可以用来设置客户端建立 Socket 连接时的超时时间。如果在 timeout 内没有成功建立连接，则抛出 TimeoutException 异常。connect(.)未设置 timeout 时，表示无限尝试建立连接没有超时。如果 timeout 的值小于 TCP 三次握手的时间，则 Socket 连接永远不会建立。

2. 读超时的设置

Socket 连接的 read 操作会阻塞当前线程，直到有新的数据到来或者超时异常产生，默认是无限等待没有超时。如果远端机器重启、远端程序崩溃或者网络断开，本端的 read 操作将一直被阻塞，因此设置 read 操作超时时间是非常重要的，否则 read 操作的当前线程可能一直被挂起。调用 socket. setSoTimeout(int timeout)可以设置 read 操作的超时时间，如果到了超时时间仍没有读到数据，read 操作会抛出一个 SocketTimeoutException。程序可以捕获这个异常，但是当前的 Socket 连接仍然是有效的。

3. 写超时的设置

当 Socket 连接的 write 操作发送数据时，如果远端机器重启、远端程序崩溃或者网络断开，TCP 模块会重传数据，若多次重传失败则自动关闭 TCP 连接。因此，Socket 连接的 write 操作超时是基于 TCP 协议栈的超时重传机制，一般不需要设置 write 操作的超时时间，也没有提供这种方法。

除了超时设置，再增加服务器的多线程处理。在高并发网络请求的环境下，服务端会用一个线程处理连接请求，当连接建立之后，该连接上的会话由另外一个线程单独处理。这样会提高服务端的响应速度。

代码 1-3：类 Server1

```
1  public abstract class Server1 {
2      private int port;
3      protected  Server1(int port){
4          this.port = port;
5      }
6      //线程池，每个会话由一个线程处理
7      private ExecutorService fixPool = Executors.newCachedThreadPool();
8      public void service(){
9          try {
10             ServerSocket serverSocket = new ServerSocket(port);
11             System.out.println("服务器开启服务端口" + port + "，等待连接请求……");
12             while(true) {
13                 Socket socket = serverSocket.accept();
14                 System.out.println("接受客户端的连接，开始会话……");
15                 //设置服务端 socket 读超时为 20000 毫秒
16                 socket.setSoTimeout(20000);
17                 //把会话操作提交给线程池，由另外的线程单独处理会话。
18                 fixPool.execute(
19                         () ->{
20                             try {
```

```
21                              session(socket);
22                              socket.close();
23                              System.out.println("会话结束,服务端关闭连接。");
24                          }catch (Exception e){
25                              e.printStackTrace();
26                          }
27                      }
28                  );
29              }
30          }catch (Exception e){
31              e.printStackTrace();
32          }
33      }
34      abstract protected void session(Socket socket) throws Exception;
35  }
```

代码 1-3 是服务端。主线程在无限 while 循环(见代码 1-3 的第 12 行)内执行第 13 行的 accept(),等待客户端的连接；每接受一个客户端的连接,就启动一个新线程来运行此连接上的会话,而主线程将再次执行 accept()等待客户端的连接。代码 1-4 是客户端,其第 13 行和第 16 行增加超时设置。

代码 1-4：类 Client1

```
1   public abstract class Client1 {
2       private String serverIP;
3       private int port;
4       protected Client1(String ip, int port){
5           this.serverIP = ip;
6           this.port = port;
7       }
8       public void communicate() {
9           InetSocketAddress isAddr = new InetSocketAddress(serverIP, port);
10          Socket socket = new Socket();
11          try {
12              //设置客户端的连接超时为 30000 毫秒
13              socket.connect(isAddr , 30000 );
14              System.out.println("客户端连接成功,开始会话……");
15              //设置客户端的读超时为 20000 毫秒
16              socket.setSoTimeout(20000);
17              session(socket);
18              socket.close();
19              System.out.println("会话结束,客户端关闭连接。");
20          }catch (Exception e){
21              e.printStackTrace();
```

```
22              }
23          }
24          abstract protected void session(Socket socket) throws Exception;
25 }
```

类Server1和类Client1均使用了“模板方法”设计模式，代表会话的抽象方法session()将在子类中得到实现。

1.3 升级write与read

发送方使用OutputStream或其子类对象的write(.)方法发送数据：

public void write(byte[] data) throws IOException

public void write(byte[] data, int offset, int length) throws IOException

而接收方使用InputStream或其子类对象的read()方法接收数据：

public int read(byte[] input) throws IOException

public int read(byte[] input, int offset, int length) throws IOException

第1个read(.)的功能是从网络通道上读取多字节填入指定的数组input，最多能读取的字节数为input的元素个数；第2个read()的功能是从网络通道上读取多字节填入指定数组input的子数组，即从offset位置开始填入，最多连续填入length字节。当执行read()而网络通道上没有传入字节时，read()方法阻塞当前线程直至有字节传入。

read(.)若返回−1则意味着InputStream的关闭，这通常是由发送方关闭对应OutputStream所致。

虽然read(.)参数指定了读入字节的存放位置及预留空间，但不是必然会接收到与空间个数等量的字节。由于网络传输的不连续性及不可预测性特点，当发送方调用write(.)方法发送了10字节、接收方调用read(.)方法接收时，即使此时read(.)的参数是10个元素的字节数组，也可能需要连续多次调用read(.)才能完整地接收到这10字节，而且期间read(.)方法可能会被阻塞。从开始发送到接收完成整个过程可能如图1-2所示。

代码1-5的函数readFully(.)可以确保接收方能完整接收到与预留空间等量的字节。

代码1-5：函数readFully(.)

```
1 private static void readFully(Socket socket, byte[ ] bytes) throws Exception{
2     InputStream ins = socket.getInputStream();
3     int bytesToRead = bytes.length;
4     int readCount = 0;
5     while (readCount < bytesToRead) {
6         int result = ins.read(bytes, readCount, bytesToRead - readCount);
```

```
7            //Stream 意外结束
8            if (result == -1) {
9                throw new Exception("异常：InputStream 意外结束!");
10           }
11           readCount += result;
12       }
13 }
```

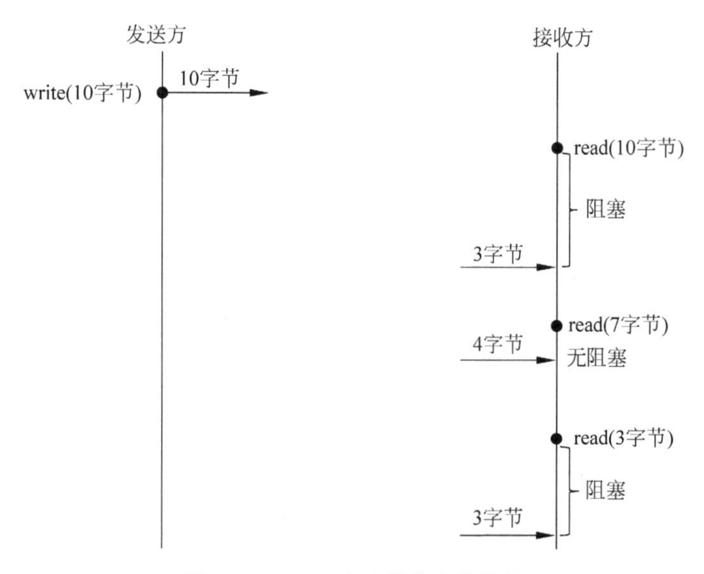

图 1-2 read(.)方法接收字节图解

虽然接收方可以为每次接收预留空间，但是接收方事先并不知道本次发送方要发送多少字节，因此发送方在发送数据之前应该告知接收方本次将要发送数据的字节总数，即接收方将要接收的字节总数，以便接收方预留等量的空间。设计代码 1-6 的函数 recv(.)，接收方首先读取表示一个整数的 4 字节，然后创建一个长度为此整数的字节数组，最后调用 readFully(.)把这个数组填满，而这个数组正是 recv(.)函数的返回值。

代码 1-6：函数 recv(.)

```
1  private static byte[] recv(Socket socket) throws Exception{
2      byte[] countBytes = new byte[4];
3      readFully(socket, countBytes);
4      int count = Utils.byteArrayToInt(countBytes);
5      byte[] dataBytes = new byte[count];
6      readFully(socket, dataBytes);
7      return dataBytes;
8  }
```

代码 1-6 的第 4 行调用了自定义的函数 byteArrayToInt(.)，该函数的功能是把字节数组转换成对应的整数，见代码 1-7。

代码 1-7：函数 byteArrayToInt(.)

```
1  public static int byteArrayToInt(byte[] b) {
2      return b[3] & 0xFF |
3              (b[2] & 0xFF) << 8 |
4              (b[1] & 0xFF) << 16 |
5              (b[0] & 0xFF) << 24;
6  }
```

代码 1-6 的 recv(.)函数是对 InputStream. read(.)方法的升级。因此，与 read(.)函数对应的 OutputStream. write(.)方法也要做相应地升级。代码 1-8 的 send(.)函数正是对 write(.)函数的升级，其功能设计为先发送一个用 4 字节表示的整数，表示随后将要发送的一组字节的字节数，之后再发送由此函数参数指定的一组字节。

代码 1-8：函数 send(.)

```
1  private static void send(Socket socket, byte[] obytes) throws Exception{
2      OutputStream outs = socket.getOutputStream();
3      int bytesCount = obytes.length;
4      //先发送字节数
5      byte[] cntBytes = Utils.intToByteArray(bytesCount);
6      outs.write(cntBytes, 0, cntBytes.length);
7      outs.flush();
8      outs.write(obytes, 0, obytes.length);
9      outs.flush();
10 }
```

代码 1-8 有两点需要关注。第一，第 7 行与第 9 行的 flush()方法，在调用 write(.)函数后调用 flush()方法，这是一个明智的做法，因为 flush()方法会强制 Stream 在其底层缓冲区未满的时候发送数据；第二，第 5 行调用的自定义函数 intToByteArray(.)，其功能是把整数转换成对应的字节数组，参见代码 1-9。

代码 1-9：函数 intToByteArray(.)

```
1  public static byte[] intToByteArray(int a) {
2      return new byte[] {
3              (byte) ((a >> 24) & 0xFF),
4              (byte) ((a >> 16) & 0xFF),
5              (byte) ((a >> 8) & 0xFF),
6              (byte) (a & 0xFF)
7      };
8  }
```

1.4 案例1：传输字符串的会话

当客户端与服务端连接成功，它们之间进行如图 1-3 所示的会话。

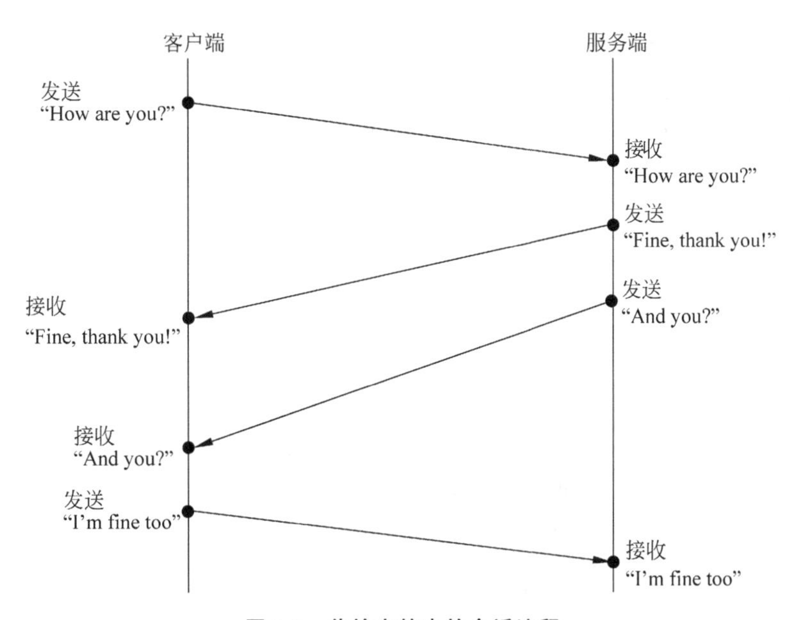

图 1-3 传输字符串的会话流程

根据该会话要求，可知双方按照如下操作时序进行通信。

客户端：发送、接收、接收、发送。

服务端：接收、发送、发送、接收。

收发时序是自定义的，这是通信协议的一部分。具体的收发内容可以变化，并不拘泥于例子。这里约定的通信格式是字符串，若能把字符串与字节数据进行相互转换，则可以调用代码 1-8 的 send(.)函数和代码 1-6 的 recv(.)函数。为此定义如下两个函数。

代码 1-10：函数 sendString(.)

```
1  public static void sendString(Socket socket, String msg) throws Exception{
2      byte[] obytes = msg.getBytes("UTF-8");
3      send(socket, obytes);
4  }
```

代码 1-11：函数 recvString(.)

```
1  public static String recvString(Socket socket) throws Exception{
2      byte[] ibytes = recv(socket);
3      String msg = new String(ibytes,0, ibytes.length,"UTF-8");
4      return msg;
5  }
```

代码1-10的sendString(.)函数在指定的socket通道上发送一个指定的字符串；而接收方则对应地调用代码1-11的recvString(.)函数在socket通道上接收到此字符串。这两个函数分别调用了send(.)和recv(.)。

现在根据会话的流程分别完善服务端和客户端。服务端参见代码1-12。

代码1-12：类Server2

```
1  public class Server2 extends Server1{
2      public static void main(String[ ] args) {
3          Server2 server = new Server2();
4          server.service();
5      }
6      public Server2(){
7          super(8001);
8      }
9      @Override
10     protected void session(Socket socket) throws Exception {
11         String msg = Receiver.recvString(socket);
12         System.out.println("Receive: " + msg);
13         msg = "Fine, thank you!";
14         Sender.sendString(socket, msg);
15         System.out.println("Send: " + msg);
16         msg = "And you?";
17         Sender.sendString(socket, msg);
18         System.out.println("Send: " + msg);
19         msg = Receiver.recvString(socket);
20         System.out.println("Receive: " + msg);
21     }
22 }
```

由于继承了Server1,Server2只需要重载session(.)。与之对应的客户端参见代码1-13。

代码1-13：类Client2

```
1  public class Client2 extends Client1{
2      public static void main(String[ ] args) {
3          ExecutorService fixPool = Executors.newCachedThreadPool();
4          int concurrentCount = 100;
5          //用多线程模拟多客户端并发通信
6          for (int i = 0;i < concurrentCount;i++) {
7              fixPool.execute(
8                      () ->{
9                          Client2 client = new Client2();
10                         client.communicate();
11                     }
12             );
```

```
13          }
14          fixPool.shutdown();
15      }
16      public Client2() {
17          super("127.0.0.1",8001);
18      }
19      @Override
20      protected void session(Socket socket) throws Exception {
21          String msg = "How are you!";
22          System.out.println("Send: " + msg);
23          Sender.sendString(socket, msg);
24          msg = Receiver.recvString(socket);
25          System.out.println("Receive: " + msg);
26          msg = Receiver.recvString(socket);
27          System.out.println("Receive: " + msg);
28          msg = "I'm fine too.";
29          Sender.sendString(socket, msg);
30          System.out.println("Send: " + msg);
31      }
32  }
```

由于继承了 Client1，Client2 只需要重载 session(.)。第 3 行增加线程池，用多线程模拟多客户端的并发访问。

把 Client2 的客户端并发个数调整为 1，先运行 Server2，再运行 Client2，服务端及客户端控制台分别显示如下信息。

服务端控制台信息	客户端控制台信息
服务器开启服务端口 8001，等待连接请求…… 接受客户端的连接，开始会话…… Receive：How are you! Send：Fine，thank you! Send：And you? Receive：I'm fine too. 会话结束，服务端关闭连接。	客户端连接成功，开始会话…… Send：How are you! Receive：Fine，thank you! Receive：And you? Send：I'm fine too. 会话结束，客户端关闭连接。

可以设置更多的客户端并发数，再次运行。

1.5 案例 2：传输对象的会话

通信双方希望在会话期间直接传递 Java 对象，会话流程如图 1-4 所示。

有如下 4 种方法，需要注意的是，接收方与发送方必须使用相同的方法。

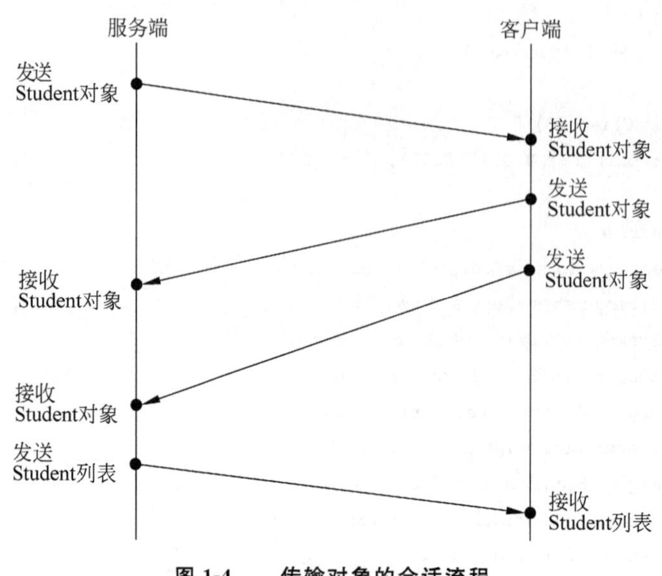

图 1-4　传输对象的会话流程

方法 1：网络通信底层传输的是字节，因此在发送时需要把 Java 对象转成字节序列再发送，在接收时需要把接收到的字节序列转成 Java 对象。这里涉及序列化和反序列化，所以传输的 Java 对象必须具有 java.io.Serializable 接口。这种方法需要自己完成序列化和反序列化操作。

方法 2：开发者直接使用 Java 的 ObjectOutputStream/ObjectInputStream 进行对象传输。其本质是 Java 已实现了方法 1，所以对象也必须具有 Serializable 接口。

方法 3：发送时把 Java 对象转成 JSON 格式的字符串，再用代码 1-10 的 sendString(.)函数发送此字符串；接收时先用代码 1-11 的 recvString(.)函数接收字符串，再把此 JSON 格式的字符串转成 Java 对象。这里的对象不必实现 Serializable 接口。

方法 4：其他的序列化方法，如把 Java 对象转成 XML 格式的字符串，再转成字节数组。

这里只列出方法 1 的代码，方法 2 和方法 3 在源代码中提供了函数。

方法 1 需要自己定义序列化操作函数和反序列化操作函数。序列化操作函数见代码 1-14，反序列化操作函数见代码 1-15。它们分别使用了 ObjectOutputStream 和 ObjectInputStream，不过并没有用它们来传输对象。

代码 1-14：函数 objSerializableToByteArray(.)

```
1  public static byte[] objSerializableToByteArray(Object objSerializable) throws Exception{
2      ByteArrayOutputStream byteArrayOutputStream = new ByteArrayOutputStream();
3      ObjectOutputStream objectOutputStream = new ObjectOutputStream(byteArrayOutputStream);
4      objectOutputStream.writeObject(objSerializable);
5      byte[] bytes = byteArrayOutputStream.toByteArray();
```

```
6       objectOutputStream.close();
7       byteArrayOutputStream.close();
8       return bytes;
9   }
```

代码 1-15：函数 byteArrayToObjSerializable(.)

```
1   public static Object byteArrayToObjSerializable(byte[] bytes) throws Exception{
2       ByteArrayInputStream byteArrayInputStream = new ByteArrayInputStream(bytes);
3       ObjectInputStream objectInputStream = new ObjectInputStream(byteArrayInputStream);
4       Object obj = objectInputStream.readObject();
5       objectInputStream.close();
6       byteArrayInputStream.close();
7       return obj;
8   }
```

在自定义的序列化和反序列化函数的基础上，再定义对象发送函数 sendObject(.)和对象接收函数 recvObject(.)，代码分别参见代码 1-16 和代码 1-17。这里分别使用了之前自定义的 send(.)函数和 recv(.)函数。

代码 1-16：函数 sendObject(.)

```
1   public static void sendObject(Socket socket, Object objSerializable) throws Exception{
2       byte[] bytes = Utils.objSerializableToByteArray(objSerializable);
3       Sender.send(socket, bytes);
4   }
```

代码 1-17：函数 recvObject(.)

```
1   public static Object recvObject(Socket socket) throws Exception{
2       byte[] bytes = Receiver.recv(socket);
3       Object obj = Utils.byteArrayToObjSerializable(bytes);
4       return obj;
5   }
```

服务端 Server3 的 session(.)函数按照以下步骤进行会话：发送对象、接收对象、接收对象、发送列表对象，见代码 1-18。例子中使用的是 Student 对象，可以替换为任何具有 Serializable 接口的对象。

代码 1-18：类 Server3

```
1   public class Server3 extends Server1{
2       public static void main(String[] args) {
3           Server3 server = new Server3();
4           server.service();
5       }
6       public Server3(){
7           super(8001);
8       }
```

```
9       @Override
10      protected void session(Socket socket) throws Exception {
11          Student stu = StuRepository.getStudent();
12          Sender.sendObject(socket, stu);
13          System.out.println("发送对象：" + stu);
14          stu = (Student)Receiver.recvObject(socket);
15          System.out.println("接收对象：" + stu);
16          stu = (Student)Receiver.recvObject(socket);
17          System.out.println("接收对象：" + stu);
18          List<Student> stuList = StuRepository.getStudents();
19          Sender.sendObject(socket, stuList);
20          System.out.println("发送列表：" + stuList);
21      }
22  }
```

通信双方的步骤必须匹配，因此与之对应的客户端 Client3 的 session(.)函数则按照以下步骤进行会话：接收对象、发送对象、发送对象、接收列表对象，见代码 1-19。

代码 1-19：类 Client3

```
1   public class Client3 extends Client1{
2       public static void main(String[] args) {
3           ExecutorService fixPool = Executors.newCachedThreadPool();
4           int concurrentCount = 50;
5           //用多线程模拟多客户端并发通信
6           for (int i = 0;i < concurrentCount;i++) {
7               fixPool.execute(
8                       () ->{
9                           Client3 client = new Client3();
10                          client.communicate();
11                      }
12              );
13          }
14          fixPool.shutdown();
15      }
16      public Client3() {
17          super("127.0.0.1",8001);
18      }
19      @Override
20      protected void session(Socket socket) throws Exception {
21          Student stu = (Student)Receiver.recvObject(socket);
22          System.out.println("接收对象：" + stu);
23          Sender.sendObject(socket, stu);
24          System.out.println("发送对象：" + stu);
25          Sender.sendObject(socket, stu);
26          System.out.println("发送对象：" + stu);
```

```
27          List<Student> stuList = (List<Student>)Receiver.recvObject(socket);
28          System.out.println("接收列表：" + stuList);
29      }
30 }
```

运行时，先启动 Server3，再启动 Client3。可以在控制台查看运行结果。

1.6 案例3：传输文件的会话

当客户端与服务端连接成功，它们之间将进行如图1-5所示的会话。

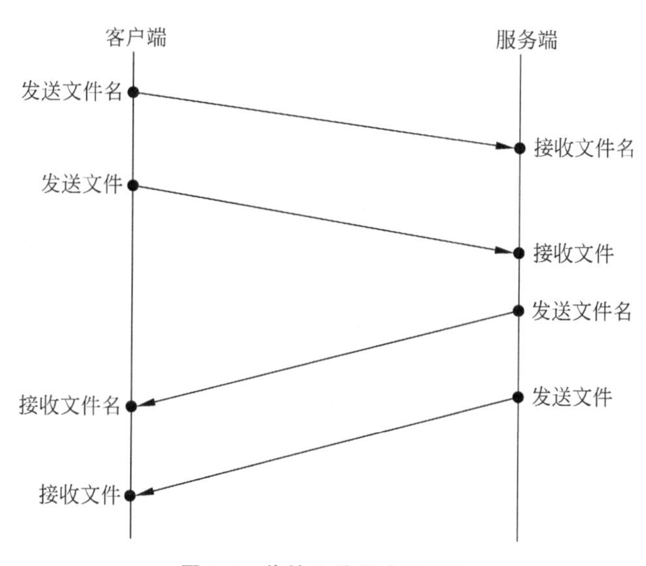

图1-5 传输文件的会话流程

服务端：接收文件名，接收文件(并保存)，发送文件名，发送文件内容。

客户端：发送文件名，发送文件内容，接收文件名，接收文件(并保存)。

字节数组的下标是整数类型，而文件长度是长整数类型，因此，对于超大文件，一次性读取文件的全部内容写入字节数组再进行发送是不可行的。必须分批读取文件内容，每次从文件中读取一组字节并转发出去，直至发送完文件的全部字节；同样地，受限于字节数组的大小，接收方也不可能一次性接收超大文件的所有内容，只能分批接收，每接收到一组字节便转存至本地文件，直至接收完远程文件的全部字节。接收方如何知道全部字节接收完毕呢？发送方在发送文件数据之前，需先发送一个表示文件总长度的长整数给接收方，接收方则以此长整数作为判断文件是否接收完毕的依据。收发双方的操作流程如图1-6所示。

图1-6中，a、b 分别表示发送方与接收方自定义的缓冲区大小，两个值可以不相等。函数 sendFile(.)实现发送文件，参见代码1-20。参数 localFileName 用来指明要发送的本地文件。

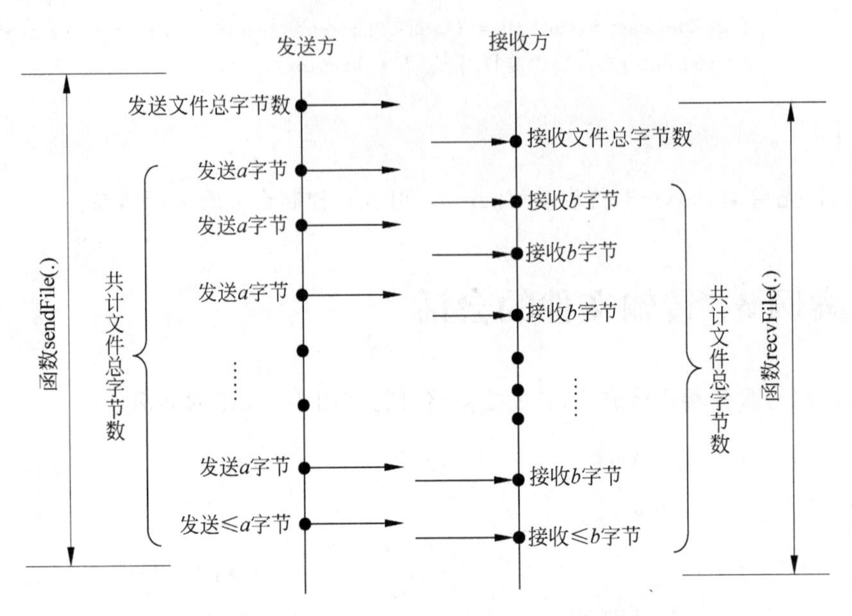

图 1-6 传输大文件的操作流程

代码 1-20：函数 sendFile(.)

```
1  public static void sendFile(Socket socket, String localFileName) throws Exception{
2      OutputStream os = socket.getOutputStream();
3      File file = new File(localFileName);
4      long fileLength = file.length();
5      //发送文件总的字节数,java 的 long 是 8 字节
6      byte[] lengthBytes = Utils.longToByteArray(fileLength);
7      os.write(lengthBytes);
8      os.flush();
9      FileInputStream fis = new FileInputStream(file);
10     //缓冲区设为 1000 * 1000 字节
11     int bufSize = 1000 * 1000;
12     byte[] buffer = new byte[bufSize];
13     //从文件读取一组字节
14     int readLength = fis.read(buffer);
15     //未到文件末尾
16     while (readLength != -1){
17         //发送一组字节
18         os.write(buffer, 0, readLength);
19         os.flush();
20         //从文件读取下一组字节
21         readLength = fis.read(buffer);
22     }
23     fis.close();
24 }
```

代码 1-20 的第 6 行调用了自定义函数 longToByteArray(.)，该函数的功能是把长整数转成对应的字节数组，参见代码 1-21。

代码 1-21：函数 longToByteArray(.)

```
1  public static byte[] longToByteArray(long x) {
2      ByteBuffer buffer = ByteBuffer.allocate(8);
3      buffer.putLong(0, x);
4      return buffer.array();
5  }
```

函数 recvFile(.)实现接收文件，参见代码 1-22。参数 savedFileName 指定了文件接收之后的保存路径。

代码 1-22：函数 recvFile(.)

```
1  public static void recvFile(Socket socket, String savedFileName) throws Exception{
2      InputStream is = socket.getInputStream();
3      FileOutputStream fos = new FileOutputStream(new File(savedFileName));
4      //先获取一个长整数的 8 字节，该长整数表示文件的长度
5      byte[] lengthBytes = new byte[8];
6      readFully(socket, lengthBytes);
7      //要接收的总的字节数
8      long restBytesToRead = Utils.byteArrayToLong(lengthBytes);
9      //缓冲区大小为 1 MB
10     int bufSize = 1024 * 1024;
11     byte[] dataBytes = new byte[bufSize];
12     do {
13         //接收最后一组字节时可能需要调整缓冲区大小
14         if (bufSize > restBytesToRead) {
15             bufSize = (int)restBytesToRead;
16             dataBytes = new byte[bufSize];
17         }
18         //填满缓冲区
19         readFully(socket, dataBytes);
20         //将缓冲区内容存入文件
21         fos.write(dataBytes);
22         //计算剩余尚未接收的字节数
23         restBytesToRead -= bufSize;
24     }while(restBytesToRead > 0);
25     fos.close();
26 }
```

代码 1-22 的第 8 行调用了自定义函数 byteArrayToLong(.)，该函数的功能是把字节数组转成对应的长整数，参见代码 1-23。

代码 1-23：函数 byteArrayToLong(.)

```
1  public static long byteArrayToLong(byte[ ] bytes) {
2      ByteBuffer buffer = ByteBuffer.allocate(8);
3      buffer.put(bytes, 0, bytes.length);
4      buffer.flip();
5      return buffer.getLong();
6  }
```

服务端 Server4 参见代码 1-24。

代码 1-24：类 Server4

```
1  public class Server4 extends Server1{
2      public static void main(String[ ] args) {
3          Server4 server = new Server4();
4          server.service();
5      }
6      public Server4(){
7          super(8001);
8      }
9      @Override
10     protected void session(Socket socket) throws Exception {
11         String fileName = Receiver.recvString(socket);
12         System.out.println("接收文件名：" + fileName);
13         String filePath = "D:\\ServerFiles\\FromClient-" + fileName;
14         Receiver.recvFile(socket, filePath);
15         System.out.println("接收并保存文件：" + filePath);
16         filePath = "D:\\ServerFiles\\xyz.mp4";
17         fileName = filePath.substring(filePath.lastIndexOf("\\") + 1);
18         Sender.sendString(socket, fileName);
19         System.out.println("发送文件名：" + fileName);
20         Sender.sendFile(socket,filePath);
21         System.out.println("发送文件：" + filePath);
22     }
23 }
```

客户端 Client4 参见代码 1-25。

代码 1-25：类 Client4

```
1  public class Client4 extends Client1{
2      public static void main(String[ ] args) {
3          ExecutorService fixPool = Executors.newCachedThreadPool();
4          //为了避免文件访问冲突,故设为 1 个客户端
5          for (int i = 0;i < 1;i++) {
6              fixPool.execute(
```

```
7                        () ->{
8                            Client4 client = new Client4();
9                            client.communicate();
10                       }
11                 );
12             }
13             fixPool.shutdown();
14         }
15         public Client4(){
16             super("127.0.0.1",8001);
17         }
18         @Override
19         protected void session(Socket socket) throws Exception {
20             String filePath = "D:\\ClientFiles\\abc.mp4";
21             String fileName = filePath.substring(filePath.lastIndexOf("\\") + 1);
22             Sender.sendString(socket, fileName);
23             System.out.println("发送文件名: " + fileName);
24             Sender.sendFile(socket,filePath);
25             System.out.println("发送文件: " + filePath);
26             fileName = Receiver.recvString(socket);
27             System.out.println("接收文件名: " + fileName);
28             filePath = "D:\\ClientFiles\\FromServer-" + fileName;
29             Receiver.recvFile(socket, filePath);
30             System.out.println("接收并保存文件: " + filePath);
31         }
32 }
```

为了避免多线程并发操作同一文件，把 Client4 的并发个数调整为 1。确保文件 D:\ServerFiles\xyz.mp4 和 D:\ClientFiles\abc.mp4 都存在，先运行 Server4，再运行 Client4，服务端及客户端控制台分别显示如下信息。

Server4 控制台显示信息	Client4 控制台显示信息
服务器开启服务端口 8001，等待连接请求…… 接受客户端的连接，开始会话…… 接收文件名：abc.mp4 接收并保存文件：D:\ServerFiles\FromClient-abc.mp4 发送文件名：xyz.mp4 发送文件：D:\ServerFiles\xyz.mp4 会话结束，服务端关闭连接	客户端连接成功，开始会话…… 发送文件名：abc.mp4 发送文件：D:\ClientFiles\abc.mp4 接收文件名：xyz.mp4 接收并保存文件：D:\ClientFiles\FromServer-xyz.mp4 会话结束，客户端关闭连接

检查对应的文件夹，可以看到新保存的文件。

习题

编写BIO程序，完成如下功能：客户端连接远程服务端，之后由使用者发送命令对服务端的文件系统进行操作，命令的顺序并不固定。

dir-列出当前文件夹内的文件及文件夹；

cd <文件夹>-改变服务端的当前文件夹；

download <远端文件名>-下载服务端的文件至本地；

upload <本地文件名>-上传本地文件至服务端；

close-结束会话。

第 2 章 NIO

NIO(non-blocking I/O)称为非阻塞式 I/O。网络 I/O 不被阻塞,故可以使用单一线程处理多个通道的数据发送与接收,以减少线程间的切换,提高运行效率。NIO 可以用在请求并发量很高的网络环境。本章对应的源代码为附件中的 nio 项目。

2.1 NIO 模型

第 1 章的 BIO 会阻塞线程,而本章的 NIO 使用选择器和 SelectableChannel 实现了非阻塞的效果。在 BIO 中无法预知通道的读写状态,因此只能采用等待即阻塞的方式直至通道就绪,然后进行数据传输;而在 NIO 中选择器是核心组件,通道向选择器注册,选择器则成了监测通道的状态是否 I/O 就绪的一种构件。这里主线程不需要以 for 循环的方式来检查每个注册通道是否有数据可以读入或者可以写出,而是操作系统底层作为通知器来通知 JVM,对应地,主线程调用选择器的 select(.)方法来监测这种就绪通知。当已注册的某个或某些通道出现了 I/O 就绪状态时,主线程则终止 select(.)方法的阻塞,继续运行,处理通道数据。这些出现就绪状态的通道也称为“被选通道”。因为主线程进行 I/O 操作时通道已经就绪,即读操作时已有数据等待读入、写操作时通道已处于可写状态,所以不会发生 I/O 阻塞,提高了 CPU 使用效率;而且使用一个线程就足以操作多个通道,也减少了 CPU 在不同的线程间非常耗时的切换动作。这种机制在 NIO 中称为“I/O 多路复用”。

图 2-1 描述了 NIO 服务端 4 个关键的部件:选择器(Selector)、通道(ServerSocketChannel 和 SocketChannel)、SelectionKey 以及单一的处理线程;每个部件各司其职,共同实现多路复用机制。

当主程序开启监听端口如 8080 时,只能绑定一个 ServerSocketChannel,因为一台机器的某个特定端口只能绑定一条通道。这种 ServerSocketChannel 只传输连接请求及应答信息,被称为连接通道。每当服务端接受客户端的一个连接请求时,一个新的 SocketChannel

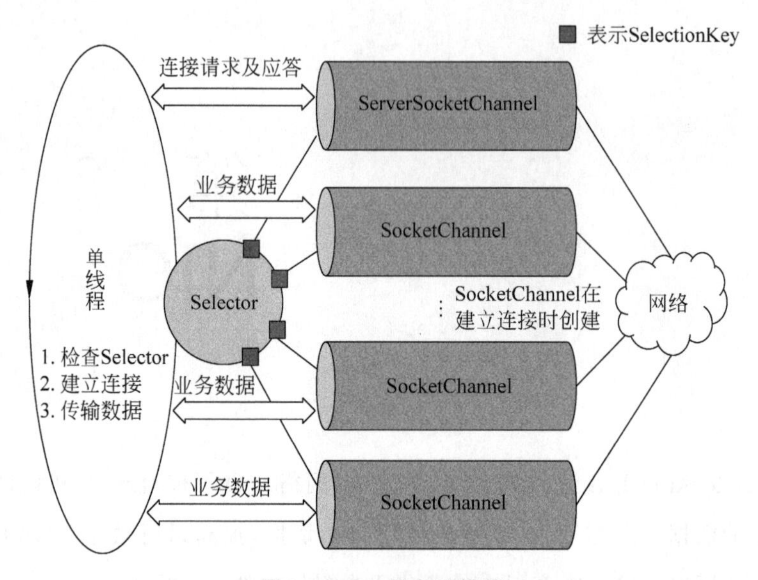

图 2-1　NIO 服务端模型

将创建。新创建的 SocketChannel 用来传输数据，被称为数据通道。不论是连接通道还是数据通道，为了便于程序监听通道的状态变化，都必须向选择器注册通道。每当一个通道完成注册时，就产生一个 SelectionKey，这个 SelectionKey 是通道注册的一个凭证，体现了通道与选择器的关联，选择器在其内部维护了一组 SelectionKey。图 2-1 中左侧是一个单线程，它的任务是调用选择器的 select(.)方法来监听已注册通道的就绪状态，当有通道出现就绪状态时，线程就进行相应的处理：针对 ServerSocketChannel 的就绪，线程的处理是接受连接请求；针对 SocketChannel 的就绪，线程的处理是读入或者写出数据。之前通道注册时产生的 SelectionKey 在这时发挥了重要作用：不但通道的就绪事件体现在 SelectionKey 里，而且线程是通过 SelectionKey 来定位通道的。

2.2　NIO 服务端框架代码

代码 2-1 体现服务端的处理流程。

代码 2-1：类 Server

```
1  public abstract class Server {
2      private Selector selector;
3      protected Server(int port) throws Exception{
4          selector = Selector.open();
5          ServerSocketChannel ssChannel = ServerSocketChannel.open();
6          ssChannel.configureBlocking(false);
7          ssChannel.bind(new InetSocketAddress("127.0.0.1", port));
8          ssChannel.register(selector, SelectionKey.OP_ACCEPT);
```

```
9           System.out.println("服务器开启服务端口" + port + ",等待连接请求……");
10      }
11      protected void closeChannel(SelectionKey key) throws Exception{
12          key.channel().close();
13          key.cancel();
14          System.out.println("会话结束,服务端关闭连接。");
15      }
16      public void service() {
17          while (true) {
18              try {
19                  selector.select();
20                  Set<SelectionKey> selectedKeysSet = selector.selectedKeys();
21                  Iterator<SelectionKey> iterator = selectedKeysSet.iterator();
22                  while (iterator.hasNext()) {
23                      SelectionKey key = iterator.next();
24                      iterator.remove();
25                      if (key.isAcceptable()) {
26                          ServerSocketChannel ssChannel =
                                        (ServerSocketChannel) key.channel();
27                          SocketChannel channel = ssChannel.accept();
28                          System.out.println("接受客户端的连接,开始会话……");
29                          channel.configureBlocking(false);
30                          registerOnConnected(channel, key.selector());
31                      }else{
32                          session(key);
33                      }
34                  }
35              } catch (Exception ex) {
36                  ex.printStackTrace();
37              }
38          }
39      }
40      abstract protected void session(SelectionKey key) throws Exception;
41      abstract protected void registerOnConnected(
                                        SocketChannel channel,
                                        Selector selector) throws Exception;
42  }
```

第3～10行的构造器代码：创建选择器；创建连接通道；把连接通道设为“非阻塞”模式，因为不论是连接通道还是数据通道，只有非阻塞模式才能使用I/O多路复用；把连接通道绑定至服务端口；向选择器注册此连接通道，注册时指明要监听的事件类型，参数SelectionKey.OP_ACCEPT表示希望选择器在此通道上只关注“接受连接就绪”事件而忽略其他类型的I/O事件。

从第 17 行开始的 while 循环就是单线程的服务功能。第 19 行的 selector. select()是查看选择器的事件通知，如果之前向选择器注册过的任一通道上发生了感兴趣的 I/O 事件(即注册时第 2 个参数指明的事件类型)，这个方法就返回，否则就一直阻塞。在高并发网络环境下，某一时刻发生 I/O 事件的通道往往不止一个，接下来调用 selector. selectedKeys()返回的就是发生感兴趣 I/O 事件的 SelectionKey 集，因为一个 SelectionKey 对应一个通道，所以这个 SelectionKey 集就意味着与之对应的多个通道中每个通道都发生了 I/O 事件，而这些事件必然是通道注册时所指定的事件类型。

从第 22 行开始的 while 循环是依次检查每个发生 I/O 事件的通道。第 24 行的 remove()调用是把事件通道从事件通道集中移除以避免下一次重复通知。随后的 if 语句是根据通道事件类型做不同的处理。

第 25 行若 key. isAcceptable()返回 true 值，则表示当前 SelectionKey 对应的通道处于“接受连接就绪”状态，第 26 行是获得与当前 SelectionKey 对应的连接通道，第 27 行是调用连接通道的 accept()方法，此方法将创建一个新的数据通道，第 29 行是把新的数据通道设置为“非阻塞”模式以便使用 I/O 多路复用特性，第 30 行是把这个数据通道注册到选择器，以便在下一轮第 19 行的 select()调用将监听到这个新数据通道上发生的 I/O 事件。

第 30 行的注册操作是一个抽象方法，因为注册时的事件类型参数必须由具体的会话协议来决定，所以具体的注册操作只能推迟到 Server 的子类里去实现。服务端的数据通道通常只需关注两类事件，SelectionKey. OP_READ 和 SelectionKey. OP_WRITE。顾名思义，SelectionKey. OP_READ 表示“读入就绪”事件，SelectionKey. OP_WRITE 表示“写出就绪”事件。会话时序规定了双方的对应式操作，服务端的接收必然对应客户端的发送，服务端的发送必然对应客户端的接收。作为非阻塞式通信，接收操作必然发生在数据通道处于“读入就绪”状态之时，而只有在注册时参数设为 SelectionKey. OP_READ 才可能监听到该通道的“读入就绪”事件；同理，发送操作必然发生在数据通道处于“写出就绪”状态之时，而只有在注册时参数设为 SelectionKey. OP_WRITE 才可能监听到该通道的“写出就绪”事件。因此 Server 的子类在实现第 30 行的 registerOnConnected(.)抽象方法时将根据会话协议决定向选择器注册的事件类型参数，如果会话是以服务端向客户端发送数据开始，则事件类型参数设为 SelectionKey. OP_WRITE；如果会话是以服务端接收来自客户端的数据开始，则事件类型参数设为 SelectionKey. OP_READ。

第 32 行调用抽象方法 session(.)，Server 的子类将重载此抽象方法实现具体的、特定的会话流程。

代码 2-1 的 Server 采用单个线程负责全部通道的操作，包括在连接通道上处理连接请求、在数据通道上处理数据的传输，在后面的章节我们会改良这种单线程的设计以提高性能。

因为服务端是单线程在操作所有通道的 I/O，所以我们可以只使用一个缓冲区（即字节数组）分时为所有的数据通道存放要发送的数据或者接收到的数据，但是这种设计显然有很大的局限，一是不便于以后扩展至多线程服务，二是不便于同一通道上数据的累次发送或者接收。因此每个数据通道需要配备一个缓冲区，这个缓冲区可以作为 SelectionKey 的 attachment 属性而存在。因为每个通道对应一个 SelectionKey，而每个 SelectionKey 具有一个 attachment 属性，这样每个通道都可以方便地找到自己的缓冲区，而且因为属性 attachment 的类型是 Object，它除了包含缓冲区还可以包含其他的信息。

2.3 NIO 客户端框架代码

客户端只需要使用一条通道。其操作过程如下：建立连接，进行会话，关闭连接。客户端框架代码参见代码 2-2。

代码 2-2：类 Client

```
1  public abstract class Client {
2      private Selector selector;
3      protected Client(String serverIP, int port) throws Exception{
4          selector = Selector.open();
5          SocketChannel channel = SocketChannel.open();
6          channel.configureBlocking(false);
7          channel.register(selector, SelectionKey.OP_CONNECT);
8          channel.connect(new InetSocketAddress(serverIP, port));
9      }
10     abstract protected void session(SelectionKey key) throws Exception;
11     abstract protected void registerOnConnected(
                                   SocketChannel channel,
                                   Selector selector) throws Exception;
12     protected void closeChannel(SelectionKey key) throws Exception{
13         key.channel().close();
14         key.selector().close();
15     }
16     public void communicate() throws Exception{
17         while (selector.isOpen()){
18             selector.select();
19             Set<SelectionKey> selectedKeysSet = selector.selectedKeys();
20             Iterator<SelectionKey> iterator = selectedKeysSet.iterator();
21             while (iterator.hasNext()){
22                 SelectionKey key = iterator.next();
23                 iterator.remove();
24                 if (key.isConnectable()){
25                     SocketChannel channel = (SocketChannel)key.channel();
26                     while(!channel.finishConnect()){ }
```

```
27                    System.out.println("客户端连接成功,开始会话……");
28                    registerOnConnected(channel, key.selector());
29                }else {
30                    session(key);
31                }
32            }
33        }
34        System.out.println("会话结束,客户端关闭连接。");
35    }
36 }
```

第 3～9 行是 Client 的构造器,依次执行如下操作：创建选择器,创建一个数据通道,把数据通道设为非阻塞模式,向选择器注册,向服务端发起连接请求。注册时第 2 个参数值为 SelectionKey.OP_CONNECT,表示选择器关注的 I/O 事件类型为"连接就绪"。当此通道处于"可连接"状态时,选择器将获得通知。

第 17 行开始的 while 循环是重复监听选择器事件,当有事件发生时将执行第 21 行的 while 循环,依次检查每个已发生事件的通道,如果当前通道处于"连接就绪"状态,则用第 26 行代码完成连接,第 28 行调用抽象方法 registerOnConnected(.),该抽象方法将在 Client 的子类中得到具体实现,其功能是把当前用作连接请求的通道重新注册到选择器。重新注册是因为随后此通道的用途不再是发送连接请求,而是进行业务数据传输,所以选择器对此通道的关注事件也不再是"连接就绪"事件,而是"读入就绪"或者"写出就绪"事件,所以要重新注册,这次注册用的事件类型参数与创建通道时注册用的事件类型参数不同。具体关注的是"读入就绪"事件还是"写出就绪"事件取决于会话流程,如果会话是以客户端发送开始,则注册时事件类型为"写出就绪"事件；如果会话是以客户端接收开始,则注册时事件类型为"读入就绪"事件,因其可变性故让子类来完成。同一个通道如果向同一个选择器多次注册,则后一次的注册会取代前一次的注册,即前一次的注册自动失效。

第 30 行调用抽象方法 session(.),Client 的子类将重载此抽象方法实现具体的、特定的会话流程。

2.4 ByteBuffer 及其在 NIO 中使用的问题

ByteBuffer 是 Java 在 NIO 中引入的类型,其本质是对字节数组的封装,为开发者提供更加方便的操作数组的方法,以提高开发效率。类似地,还有 CharBuffer、DoubleBuffer、FloatBuffer、IntBuffer、LongBuffer、ShortBuffer,从名字上就可以看出,它们是对相应基本类型数组的封装,如 IntBuffer 是对 int 数组的封装。在网络应用开发中主要使用 ByteBuffer。

ByteBuffer 提供的操作本质上是对数组的操作。在创建一个 ByteBuffer 对象时可以使用其静态方法 ByterBuffer. allocate(.)或者对一个无效内容的字节数组调用静态方法 ByteBuffer. wrap(.)，所创建的 ByteBuffer 对象处于“写模式”，等待程序填入元素。写模式时的 ByteBuffer 对象可调用的操作包括 ByteBuffer 的 put(.)方法以及 SocketChannel 的 read(.)方法；对一个有效内容的字节数组调用静态方法 ByteBuffer. wrap(.)也会创建一个 ByteBuffer 对象，该对象处于“读模式”，等待程序读取元素，此时可调用的操作包括 ByteBuffer 的 get(.)方法以及 SocketChannel 的 write(.)方法。处于写模式的 ByteBuffer 对象调用 flip()方法，此 ByteBuffer 将转成读模式；处于读模式的 ByteBuffer 对象调用 clear()方法，此 ByteBuffer 将转成写模式。不论 ByteBuffer 处于何种模式，调用 ByteBuffer 的 hasRemaining()方法可以知道此 ByteBuffer 是否还有(写模式时的)空间或者(读模式时的)数据可用；调用 ByteBuffer 的 remaining()方法则可以知道此 ByteBuffer 还有多少个(写模式时的)空间字节或者多少个(读模式时的)数据字节。图 2-2 是 ByteBuffer 的状态转换图。

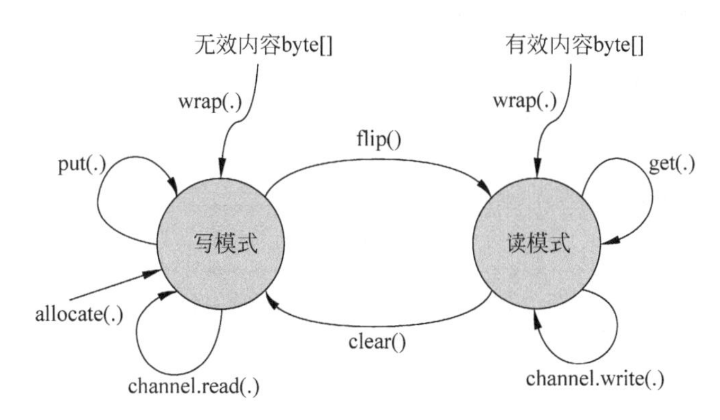

图 2-2 ByteBuffer 的状态转换图

现在来看 ByteBuffer 在网络 I/O 中的使用。

假设有一个 ByteBuffer，现在我们要求从数据通道接收数据把它填满。初步的代码可能是如下这样的。

```
ByteBuffer buf = ByteBuffer.allocate(1024);
...
if (key.isReadable()){
    SocketChannel channel = (SocketChannel) key.channel();
    channel.read(buf);
}
```

然而这个代码并不能奏效。虽然我们是在通道处于“读入就绪”状态下调用 channel. read(buf)，但是并不能保证把 buf 填满，因为 NIO 通过选择器改善了对通道的管理，但并没有改变 TCP 特有的流式字节传输特征。那么,采用如下代码能否保证把 buf 填满呢?

```
if (key.isReadable()){
    SocketChannel channel = (SocketChannel) key.channel();
    while(buf.hasRemaining()){
        channel.read(buf);
    }
}
```

这里使用了 while 循环,确实可以把 buf 填满,但是仔细分析,就会发现这段代码存在这样的问题:channel.read(buf)的第一次调用为什么不能填满 buf?因为传入通道的数据不够多。第一次调用 read(.)已经把通道内现有的数据全部取出,随后 while 循环的第二次调用 read(.)就一定能取到数据吗?这很难说,因为很难保证数据正在源源不断地传入通道。所以 while 循环体内的 read(.)可能出现取不到数据的空读入,这显然降低了 I/O 效率;更不妙的是,这种循环检查的方式与阻塞式读入并无两样,这完全违背了 NIO"非阻塞"的初衷。这个问题将在下一章得到解决。

数据写出是不是也面临着同样的问题呢?

假设有一个 ByteBuffer,里面已经填满数据,现在我们要求从数据通道把这些数据全部发送出去,初步的代码可能是如下这样的。

```
ByteBuffer buf = ByteBuffer.wrap(bytes);
...
if (key.isWritable()){
    SocketChannel channel = (SocketChannel) key.channel();
    channel.write(buf);
}
```

同样,这个代码并不能奏效。由于网络底层缓冲区等原因,channel.write(buf)的一次调用不能确保参数 buf 的数据全部发送出去,也就是说,在 buf 里还有遗留数据尚未发送。那么采用如下代码能否保证把 buf 的数据全部发送出去呢?

```
if (key.isWritable()){
    SocketChannel channel = (SocketChannel) key.channel();
    while(buf.hasRemaining()){
        channel.write(buf);
    }
}
```

增加 while 循环确实可以把 buf 的数据全部发出去,但是循环体在第二次执行时并不能保证数据通道是处于"写出就绪"状态的,因此这个循环其实是在等待通道"写出就绪"状态的出现。在通道的下一个"写出就绪"状态出现之前,channel.write(.)的执行就是零数据发送的空写出,这样不但降低了 I/O 效率,而且违背了 NIO"非阻塞"的初衷。这个问题将在2.5节得到解决。

2.5 NIO的分帧处理

会话是一个由发送和接收操作组成的序列，因此需要知道本次发送或者接收在会话中的步骤号。另外，TCP传输的流式特征依然存在，因此接收方无法正确地分帧，数据帧在这里指一段逻辑意义完整的数据。例如，发送方发送一个字符串，这个字符串转换成的字节序列就是一帧。由于网络延迟等因素，这个数据帧并不是一次性完整地到达接收方，而是不规则地分段到达，因此接收方并不知道何时结束这次接收操作；再如，发送方连续发送两次，第一次发送字符串A，第二次发送字符串B，这两帧数据通过网络到达接收方，接收方最后接收到的就是一组字节，接收方并不知道帧的分界点在哪里，所以无法正确地还原出字符串A和字符串B。上述两个问题的症结在于接收方事先不知道帧的长度。为了接收方可以正确地分帧，发送方需要在发送每个数据帧之前先发送一个整数(4字节)给接收方，接收方根据首先收到的这个整数就可以对后续收到的字节序列进行正确地分帧。基于以上的考虑，设计如图2-3的类来进行对象的传输。

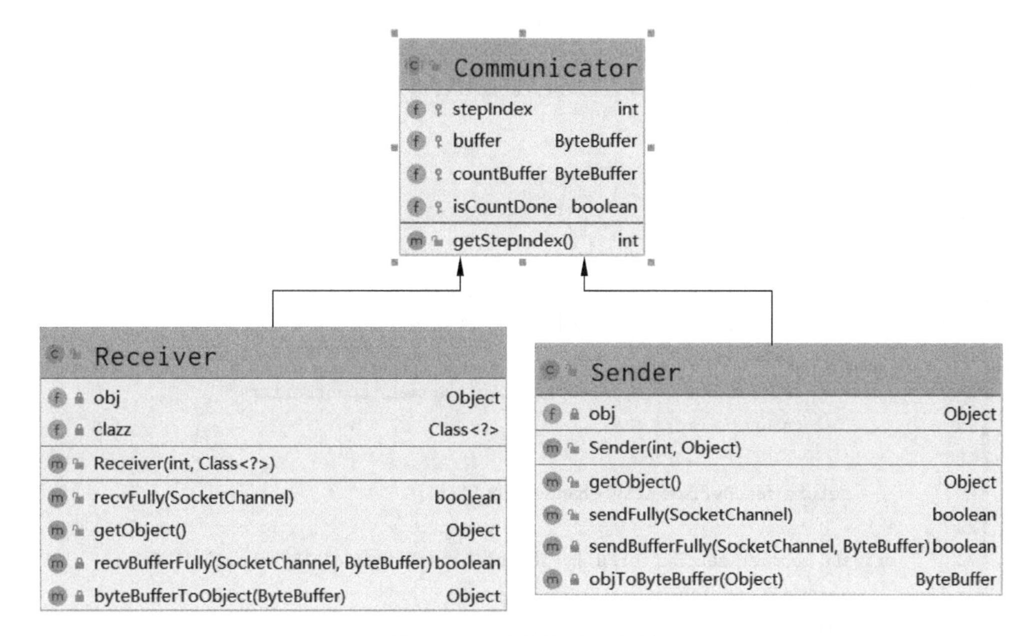

图2-3 用于传输对象的类

基类Communicator只定义了一些成员变量，见代码2-3。

代码2-3：类Communicator

```
1  abstract public class Communicator {
2       protected int stepIndex;
```

```
3        protected ByteBuffer buffer;
4        protected ByteBuffer countBuffer;
5        protected boolean isCountDone = false;
6        public int getStepIndex() {
7            return stepIndex;
8        }
9    }
```

代码 2-4 的 Sender 封装了给数据帧增加一个表示其字节数的头部信息这个操作。使用者只需要在构造时传入要发送的对象，发送时调用 sendFully(.)，若该方法返回 true 则表示已经完整地发送该对象，若该方法返回 false 则表示尚未完成对一个对象的全部发送。sendFully(.)方法是非阻塞的，解决了 2.4 节提出的问题。

代码 2-4：类 Sender

```
1   public class Sender extends Communicator {
2       //要发送的对象
3       private Object obj;
4       public Sender(int i, Object object) throws Exception{
5           this.obj = object;
6           //对象 -> ByteBuffer
7           buffer = objToByteBuffer(obj);
8           countBuffer = ByteBuffer.allocate(4);
9           countBuffer.putInt(0, buffer.capacity());
10          this.stepIndex = i;
11      }
12      public Object getObject() {
13          return obj;
14      }
15      //若全部发送，则返回 true；则返回 false，有待下一次继续发送
16      public boolean sendFully(SocketChannel channel) throws Exception{
17          if (!isCountDone){
18              isCountDone = sendBufferFully(channel, countBuffer);
19              return false;
20          }
21          return sendBufferFully(channel, buffer);
22      }
23      private boolean sendBufferFully(SocketChannel channel, ByteBuffer buf)
            throws Exception{
24          channel.write(buf);
25          return !buf.hasRemaining();
26      }
27      private ByteBuffer objToByteBuffer(Object object) throws Exception{
28          //字符串的序列化不同于一般对象的序列化
29          if (object instanceof String){
30              String st = (String)object;
31              return ByteBuffer.wrap(st.getBytes("UTF-8"));
```

```
32          }
33          if (object instanceof Serializable){
34              return ByteBuffer.wrap(Utils.objSerializableToByteArray(object));
35          }
36          throw new Exception("没有序列化接口,无法转成 ByteBuffer!");
37      }
38 }
```

代码 2-4 的第 27～37 行是把要发送的对象转成 ByteBuffer。如果对象是字符串，则直接转成字节数组再封装到 ByteBuffer 中；如果对象是其他类型的对象但是具有 Serializable 接口，则采用 Java 序列化的方式，函数 objSerializableToByteArray(.)的定义见代码 1-14；如果对象不具有 Serializable 接口，则抛出异常。序列化方式并不限于这两种，可以自定义其他的序列化方式。

代码 2-5 的 Receiver 封装了在读取数据帧之前先读取表示其字节个数的 4 字节这个操作。使用者只需要在接收时调用 recvFully(.)，若该方法返回 true 则表示已经读取一个完整的对象，若该方法返回 false 则表示尚未完成对一个对象的全部读取。recvFully(.)方法是非阻塞的，解决了 2.4 节提出的问题。

代码 2-5：类 Receiver

```
1  public class Receiver extends Communicator {
2      //存储所接收的对象
3      private Object obj;
4      private Class <?> clazz;
5      public Receiver(int i, Class <?> clazz){
6          this.clazz = clazz;
7          countBuffer = ByteBuffer.allocate(4);
8          buffer = null;
9          obj = null;
10         this.stepIndex = i;
11     }
12     public boolean recvFully(SocketChannel channel) throws Exception{
13         if (!isCountDone){
14             isCountDone = recvBufferFully(channel, countBuffer);
15             return false;
16         }
17         if (buffer == null){
18             buffer = ByteBuffer.allocate(countBuffer.getInt(0));
19         }
20         return recvBufferFully(channel, buffer);
21     }
22     public Object getObject() throws Exception{
23         //将 ByteBuffer 内部的 byte[]反序列化得到对象
24         if (!buffer.hasRemaining()){
```

```
25              buffer.flip();
26              obj = byteBufferToObject(buffer);
27          }
28          return obj;
29      }
30      private boolean recvBufferFully(SocketChannel channel, ByteBuffer buf) throws Exception{
31          channel.read(buf);
32          return !buf.hasRemaining();
33      }
34      private Object byteBufferToObject(ByteBuffer buffer) throws Exception{
35          byte[] bytes = buffer.array();
36          if (clazz.equals(String.class)){
37              return new String(bytes,"UTF-8");
38          }
39          return Utils.byteArrayToObjSerializable(bytes);
40      }
41  }
```

代码 2-5 的第 34～40 行是把接收到的 ByteBuffer 转换成对象。如果目标对象是字符串，则从 ByteBuffer 里取出字节数组直接构造字符串；如果目标对象是其他类型，则采用 Java 反序列化的方式，函数 byteArrayToObjSerializable(.)见代码 1-15。反序列化方式并不限于这两种，我们可以对应自定义的序列化方式设计自定义的反序列化方式。

2.6 案例1：传输字符串的会话

现在使用 NIO 完成如图 2-4 所示的会话。

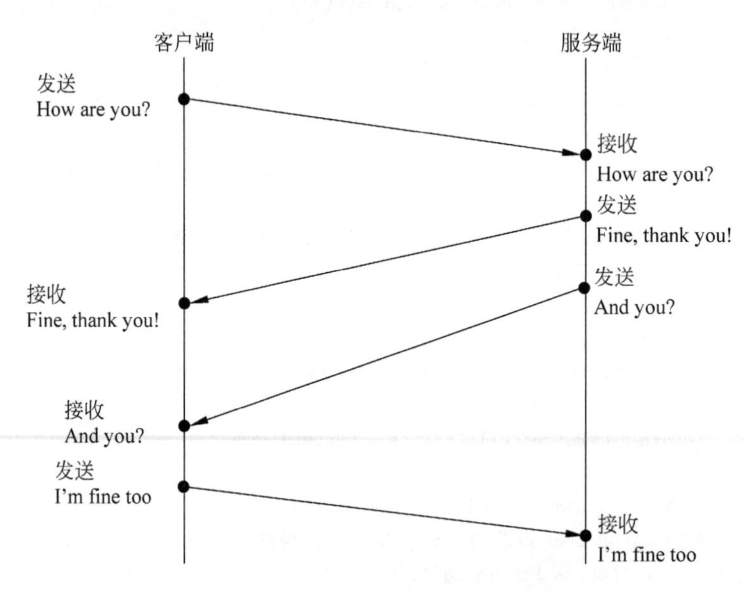

图 2-4 使用字符串进行会话的流程

服务端代码见代码 2-6，对类 Server 进行扩展。

代码 2-6：类 Server1

```
1  public class Server1 extends Server{
2      public static void main(String[] args) throws Exception{
3          Server1 server = new Server1();
4          server.service();
5      }
6      public Server1() throws Exception{
7          super(8001);
8      }
9      @Override
10     protected void registerOnConnected(SocketChannel channel, Selector selector)
           throws ClosedChannelException {
11         channel.register(selector,
                       SelectionKey.OP_READ,
                       new Receiver(0, String.class));
12     }
13     @Override
14     protected void session(SelectionKey key) throws Exception{
15         SocketChannel channel = (SocketChannel) key.channel();
16         if (key.isReadable()){
17             Receiver receiver = (Receiver)key.attachment();
18             if (receiver.recvFully(channel)){
19                 String s = (String)receiver.getObject();
20                 System.out.println("Receive: " + s);
21                 if (receiver.getStepIndex() == 0) {
22                     channel.register(key.selector(),
                       SelectionKey.OP_WRITE,
                       new Sender(receiver.getStepIndex() + 1,"I'm Fine, thank u!"));
23                 }
24                 if (receiver.getStepIndex() == 3){
25                     this.closeChannel(key);
26                 }
27             }
28         }else{  //key.isWritable()
29             Sender sender = (Sender)key.attachment();
30             if (sender.sendFully(channel)){
31                 System.out.println("Send: " + sender.getObject());
32                 //根据情况设定 Sender 的信息
33                 if (sender.getStepIndex() == 1) {
34                     //关注的事件类型不变，故不用重复注册，只需设置 attachment
35                     key.attach(new Sender(sender.getStepIndex() + 1, "And u?"));
36                 }
37                 if (sender.getStepIndex() == 2) {
```

```
38                      channel.register(key.selector(),
                        SelectionKey.OP_READ,
                        new Receiver(sender.getStepIndex() + 1, String.class));
39                  }
40              }
41          }
42      }
43 }
```

Server1 对 Server 的 session(.)和 registerOnConnected(.)进行了重载。因为会话要求当双方建立连接后客户端发送而服务端接收,所以第 11 行 register(.)的第 2 个参数是 SelectionKey. OP_READ,第 3 个参数是 Receiver 对象,用于执行接收操作。

客户端代码见代码 2-7,对类 Client 进行扩展。

代码 2-7:类 Client1

```
1  public class Client1 extends Client{
2      public static void main(String[] args) throws Exception{
3          ExecutorService fixPool = Executors.newCachedThreadPool();
4          int concurrentCount = 3;
5          for (int i = 0;i < concurrentCount;i++) {
6              fixPool.execute(
7                      () ->{
8                          try {
9                              Client1 client = new Client1();
10                             client.communicate();
11                         }catch (Exception ex){
12                             ex.printStackTrace();
13                         }
14                     }
15             );
16         }
17         fixPool.shutdown();
18     }
19     public Client1() throws Exception{
20         super("127.0.0.1",8001);
21     }
22     @Override
23     protected void registerOnConnected(SocketChannel channel, Selector selector)
           throws Exception {
24         channel.register(selector,
                           SelectionKey.OP_WRITE,
                           new Sender(0, "How are you?"));
25     }
26     @Override
```

```
27      protected void session(SelectionKey key) throws Exception {
28          SocketChannel channel = (SocketChannel)key.channel();
29          if (key.isReadable()) {
30              Receiver receiver = (Receiver)key.attachment();
31              if (receiver.recvFully(channel)){
32                  String s = (String)receiver.getObject();
33                  System.out.println("Receive: " + s);
34                  if (receiver.getStepIndex() == 1){
35                      //关注的事件类型不变,故不用重复注册,只需设置 attachment
36                      key.attach(
                            new Receiver(receiver.getStepIndex() + 1,String.class));
37                  }
38                  if (receiver.getStepIndex() == 2) {
39                      channel.register(key.selector(),
                        SelectionKey.OP_WRITE,
                        new Sender(receiver.getStepIndex() + 1, "I'm fine, too."));
40                  }
41              }
42          }else {
43              Sender sender = (Sender)key.attachment();
44              if (sender.sendFully(channel)){
45                  System.out.println("Send: " + sender.getObject());
46                  if (sender.getStepIndex() == 0) {
47                      channel.register(key.selector(),
                        SelectionKey.OP_READ,
                        new Receiver(sender.getStepIndex() + 1, String.class));
48                  }
49                  if (sender.getStepIndex() == 3) {
50                      this.closeChannel(key);
51                  }
52              }
53          }
54      }
55  }
```

Client1 对 Client 的 session(.)和 registerOnConnected(.)进行了重载。因为会话要求当双方建立连接后客户端发送而服务端接收,所以第 24 行 register(.)的第 2 个参数是 SelectionKey. OP_WRITE,第 3 个参数是 Sender 对象,用于执行发送操作。

运行时先启动附件代码 nio 项目的 Server1,再启动附件代码 nio 项目的 Client1。

2.7 案例 2: 传输对象的会话

由于 Sender 和 Receiver 针对的是任何对象,所以进行使用对象的会话也很容易。

代码 2-8 定义了具有 Serializable 接口的实体类 Student 和 Book,每个类保存于一个单独的文件。

代码 2-8：两个实体类的定义

```
1  @Data
2  @AllArgsConstructor
3  @NoArgsConstructor
4  public class Student implements Serializable {
5      Integer id;
6      String name;
7      Sex gender;
8      LocalDate birthday;
9      double gpa;
10     List<Book> reads;
11 }
12
13 @Data
14 @AllArgsConstructor
15 public class Book implements Serializable {
16     String name;
17     String author;
18     Double price;
19     int year;
20     Booktype type;
21 }
```

会话流程设计如图 2-5 所示。

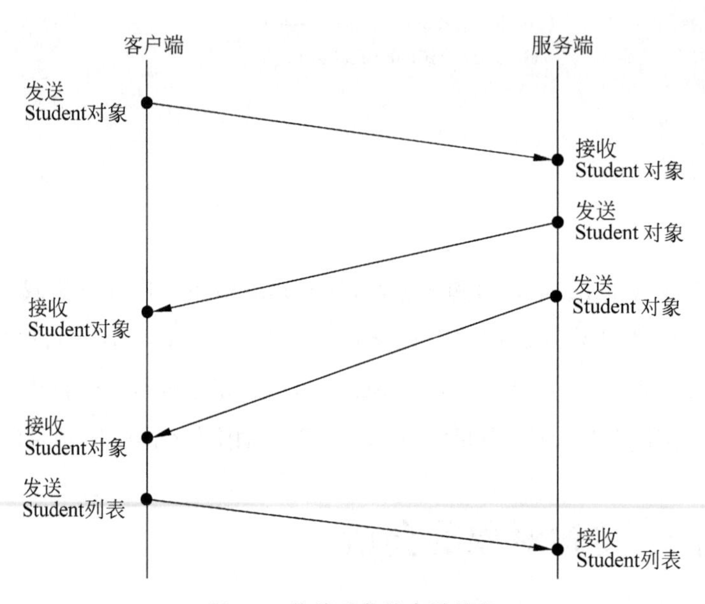

图 2-5　传输对象的会话流程

服务端代码见代码 2-9，对类 Server 进行扩展。

代码 2-9：类 Server2

```
1  public class Server2 extends Server{
2      public static void main(String[ ] args) throws Exception{
3          Server2 server = new Server2();
4          server.service();
5      }
6      public Server2() throws Exception{
7          super(8001);
8      }
9      @Override
10     protected void registerOnConnected(SocketChannel channel, Selector selector)
       throws Exception {
11         channel.register(selector,
                           SelectionKey.OP_READ,
                           new Receiver(0, Student.class));
12     }
13     @Override
14     protected void session(SelectionKey key) throws Exception {
15         //服务端会话操作：读、写、写、读(列表)
16         SocketChannel channel = (SocketChannel) key.channel();
17         if (key.isReadable()) {
18             Receiver receiver = (Receiver)key.attachment();
19             if (receiver.recvFully(channel)){
20                 Object obj = receiver.getObject();
21                 if (obj.getClass().equals(Student.class)) {
22                     System.out.println("接收单个学生：" + obj);
23                 }
24                 if (obj instanceof List){
25                     System.out.println("接收学生列表：" + obj);
26                 }
27                 if (receiver.getStepIndex() == 0) {
28                     channel.register(key.selector(),
                                        SelectionKey.OP_WRITE,
                                        new Sender(receiver.getStepIndex() + 1,
                                                   StuRepository.getStudent()));
29                 }
30                 if (receiver.getStepIndex() == 3){
31                     this.closeChannel(key);
32                 }
33             }
34         }else {
35             Sender sender = (Sender)key.attachment();
36             if (sender.sendFully(channel)){
```

```
37                  System.out.println("发送单个学生：" + sender.getObject());
38                  //根据情况设定 Sender 的信息
39                  if (sender.getStepIndex() == 1) {
40                      //因为允许的操作相同,只需修改 attachment 对象
41                      key.attach(new Sender(sender.getStepIndex() + 1,
                                   StuRepository.getStudent()));
42                  }
43                  if (sender.getStepIndex() == 2) {
44                      channel.register(key.selector(),
                                          SelectionKey.OP_READ,
                                          new Receiver(sender.getStepIndex() + 1,
                                                       List.class));
45                  }
46              }
47          }
48      }
49  }
```

Server2 对 Server 的 session(.)和 registerOnConnected(.)进行了重载。因为会话要求当双方建立连接后客户端发送而服务端接收，所以第 11 行 register(.)的第 2 个参数是 SelectionKey. OP_READ，第 3 个参数是 Receiver 对象，用于执行接收操作。

客户端代码见代码 2-10，对类 Client 进行扩展。

代码 2-10：类 Client2

```
1   public class Client2 extends Client{
2       public static void main(String[] args) throws Exception{
3           ExecutorService fixPool = Executors.newCachedThreadPool();
4           int concurrentCount = 3;
5           for (int i = 0;i < concurrentCount;i++) {
6               fixPool.execute(
7                       () ->{
8                           try {
9                               Client2 client = new Client2();
10                              client.communicate();
11                          }catch (Exception ex){
12                              ex.printStackTrace();
13                          }
14                      }
15              );
16          }
17          fixPool.shutdown();
18      }
19      public Client2() throws Exception{
20          super("127.0.0.1",8001);
```

```
21      }
22      @Override
23      protected void registerOnConnected(SocketChannel channel, Selector selector)
        throws Exception {
24          Student student = StuRepository.getStudent();
25          channel.register(selector, SelectionKey.OP_WRITE, new Sender(0,student));
26      }
27      @Override
28      protected void session(SelectionKey key) throws Exception {
29          //客户端会话操作：写、读、读、写(列表)
30          SocketChannel channel = (SocketChannel) key.channel();
31          if (key.isReadable()) {
32              Receiver receiver = (Receiver)key.attachment();
33              if (receiver.recvFully(channel)){
34                  System.out.println("接收单个学生：" + receiver.getObject());
35                  if (receiver.getStepIndex() == 1){
36                      //不用重复注册,只需修改 attachment
37                      key.attach(new Receiver(receiver.getStepIndex() + 1,Object.class));
38                  }
39                  if (receiver.getStepIndex() == 2) {
40                      channel.register(key.selector(),
                                    SelectionKey.OP_WRITE,
                                    new Sender(receiver.getStepIndex() + 1,
                                            StuRepository.getStudents()));
41                  }
42              }
43          } else{
44              Sender sender = (Sender) key.attachment();
45              if (sender.sendFully(channel)){
46                  if (sender.getObject().getClass().equals(Student.class)){
47                      System.out.println("发送单个学生：" + sender.getObject());
48                  }
49                  if (sender.getObject() instanceof List){
50                      System.out.println("发送学生列表：" + sender.getObject());
51                  }
52                  if (sender.getStepIndex() == 0) {
53                      channel.register(key.selector(),
                                    SelectionKey.OP_READ,
                                    new Receiver(sender.getStepIndex() + 1,
                                            Object.class));
54                  }
55                  if (sender.getStepIndex() == 3) {
56                      this.closeChannel(key);
57                  }
58              }
```

```
59            }
60        }
61 }
```

Client2 对 Client 的 session(.)和 registerOnConnected(.)进行了重载。因为会话要求当双方建立连接后客户端发送而服务端接收，所以第 25 行 register(.)的第 2 个参数是 SelectionKey. OP_WRITE，第 3 个参数是 Sender 对象，用于执行发送操作。

运行时先启动附件代码 nio 项目的 Server2，再启动附件代码 nio 项目的 Client2。

2.8 案例 3：传输文件的会话

传输大文件时，一次性读取文件的全部内容再写入 ByteBuffer 进行发送是不可行的。我们必须分批读取文件内容，一旦 ByteBuffer 被写满就转发出去，直至发送完文件的全部字节；同样，受限于 ByteBuffer 的最大容量，接收方也不可能一次性接收超大文件的所有内容，只能分批地接收，一旦接收的字节把 ByteBuffer 填满就转存至本地文件，直至接收完远程文件的全部字节。接收方如何知道全部字节接收完毕呢？发送方在发送文件数据之前，先发送一个表示文件总长度的长整数(8 字节)给接收方，接收方则以此长整数作为判断文件是否接收完毕的依据。收发双方的操作流程如图 2-6 所示。

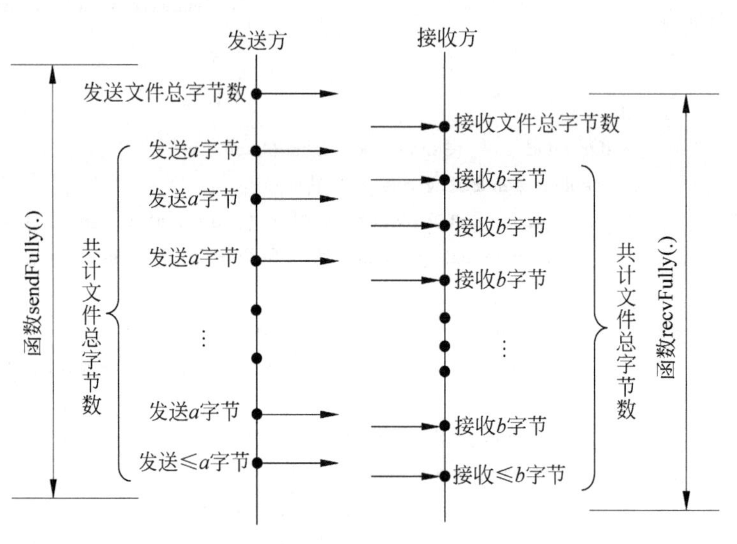

图 2-6 传输大文件的处理流程

可见 NIO 传输超大文件的思路与 BIO 的思路是一样的，但是实现细节不同。定义 Communicator 的两个子类 FileSender 和 FileReceiver，用来封装文件的传输，见图 2-7。

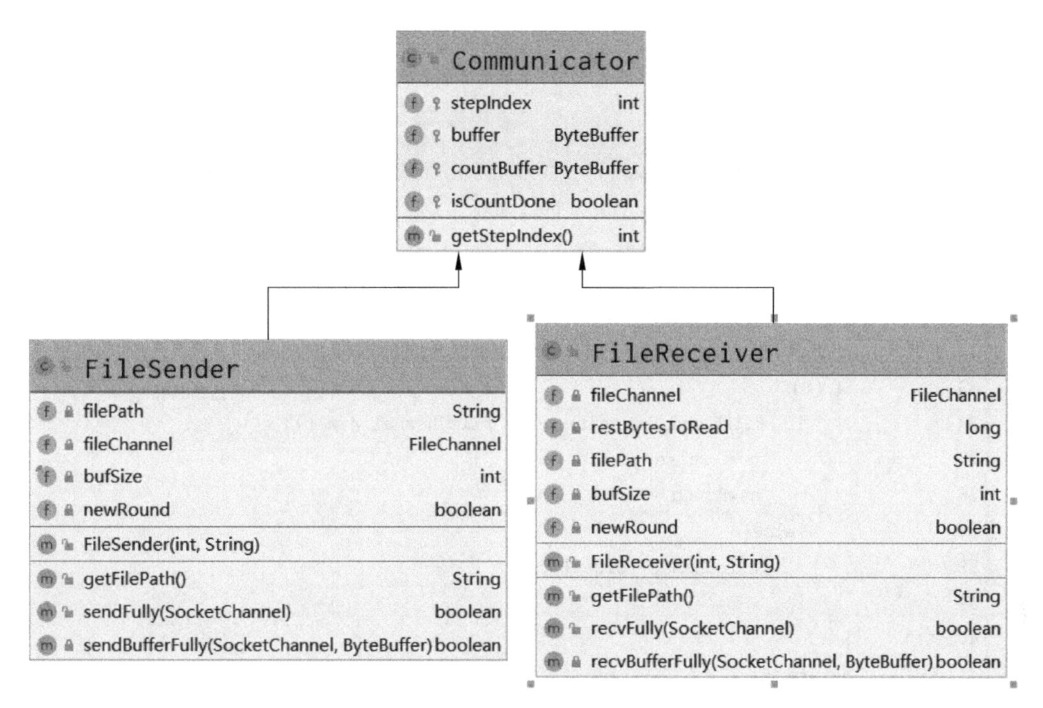

图 2-7 传输文件的类

代码 2-11：类 FileSender

```
1  public class FileSender extends Communicator{
2      private String filePath;
3      private FileChannel fileChannel;
4      private final int bufSize = 2000 * 2000;
5      private boolean newRound;
6      public FileSender(int i, String filePath) throws Exception{
7          this.filePath = filePath;
8          File file = new File(filePath);
9          FileInputStream fis = new FileInputStream(file);
10         fileChannel = fis.getChannel();
11         buffer = ByteBuffer.allocate(bufSize);
12         countBuffer = ByteBuffer.allocate(8);
13         countBuffer.putLong(0, file.length());
14         newRound = true;
15         this.stepIndex = i;
16     }
17     public String getFilePath() {
18         return filePath;
19     }
20     //若全部发送,则返回 true; 否则返回 false,有待下一次继续发送
21     public boolean sendFully(SocketChannel channel) throws Exception{
22         if (!isCountDone){
```

```
23              isCountDone = sendBufferFully(channel, countBuffer);
24              return false;
25          }
26          if (newRound){
27              this.buffer.clear();                    //将 buffer 设为写模式
28              fileChannel.read(this.buffer);
29              newRound = false;
30              this.buffer.flip();                     //将 buffer 设为读模式
31          }
32          boolean t = sendBufferFully(channel, buffer);
33          if (t) {
34              if (fileChannel.position() < fileChannel.size()) {
35                  t = false;
36                  newRound = true;
37              }else{
38                  fileChannel.close();
39              }
40          }
41          return t;
42      }
43      private boolean sendBufferFully(SocketChannel channel, ByteBuffer buf)
        throws Exception{
44          channel.write(buf);
45          return !buf.hasRemaining();
46      }
47  }
```

类 FileSender 封装了在发送文件数据之前增加一个表示文件长度的头部信息这个操作。使用者只需要在构造 FileSender 对象时传入文件路径，发送时调用 sendFully(.)，若该方法返回 true 则表示已经完整地发送该文件，若该方法返回 false 则表示尚未完成对这个文件的全部发送。sendFully(.)方法以非阻塞的方式完成文件内容的全部发送。

代码 2-12：类 FileReceiver

```
1  public class FileReceiver extends Communicator{
2      private FileChannel fileChannel;
3      private long restBytesToRead;
4      private String filePath;
5      private int bufSize = 2048 * 2;
6      private boolean newRound;
7      public FileReceiver(int i, String filePath) throws Exception{
8          this.filePath = filePath;
9          File file = new File(filePath);
10         FileOutputStream fos = new FileOutputStream(file);
11         fileChannel = fos.getChannel();
```

```
12          countBuffer = ByteBuffer.allocate(8);
13          buffer = ByteBuffer.allocate(bufSize);
14          this.filePath = filePath;
15          this.stepIndex = i;
16          newRound = true;
17      }
18      public String getFilePath() {
19          return filePath;
20      }
21      public boolean recvFully(SocketChannel channel) throws Exception{
22          if (!isCountDone){
23              isCountDone = recvBufferFully(channel, countBuffer);
24              if (isCountDone){
25                  restBytesToRead = countBuffer.getLong(0);
26              }
27              return false;
28          }
29          if (newRound) {
30              buffer.clear();           //写模式
31              newRound = false;
32              if (bufSize > restBytesToRead) {
33                  bufSize = (int) restBytesToRead;
34                  buffer = ByteBuffer.allocate(bufSize);
35              }
36          }
37          boolean t = recvBufferFully(channel, buffer);
38          if (t){
39              buffer.flip();            //读模式
40              while (buffer.hasRemaining()) {
41                  fileChannel.write(buffer);
42              }
43              restBytesToRead -= buffer.capacity();
44              if (restBytesToRead > 0){
45                  t = false;
46                  newRound = true;
47              }else{
48                  fileChannel.close();
49              }
50          }
51          return  t;
52      }
53      private boolean recvBufferFully(SocketChannel channel, ByteBuffer buf)
        throws Exception{
54          channel.read(buf);
55          return !buf.hasRemaining();
56      }
57  }
```

类 FileReceiver 封装了在接收文件数据之前先接收表示文件长度的 8 字节这个操作。使用者只需要在接收时调用 recvFully(.)，若该方法返回 true 则表示已经接收文件全部的数据，若该方法返回 false 则表示尚未接收到文件的全部数据。recvFully(.)方法以非阻塞的方式完成对文件的全部接收。

文件传输的会话流程设计如图 2-8 所示。

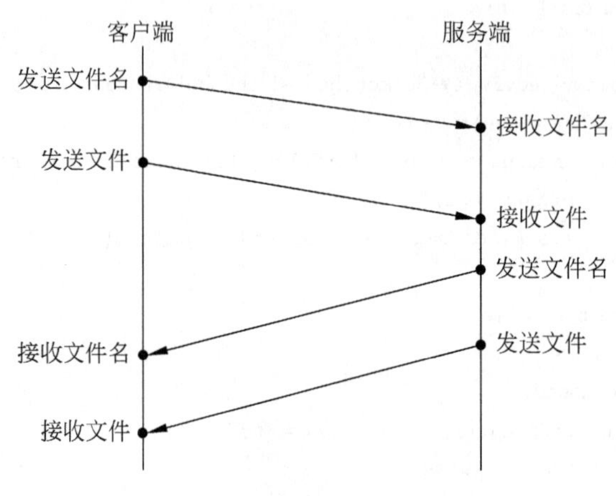

图 2-8　传输文件的会话流程

服务端代码见代码 2-13，对类 Server 进行扩展。

代码 2-13：类 Server3

```
1  public class Server3 extends Server{
2      public static void main(String[ ] args) throws Exception{
3          Server3 server = new Server3();
4          server.service();
5      }
6      public Server3() throws Exception{
7          super(8001);
8      }
9      @Override
10     protected void registerOnConnected(SocketChannel channel, Selector selector)
       throws Exception {
11         channel.register(selector,
                               SelectionKey.OP_READ,
                               new Receiver(0, String.class));
12     }
13     @Override
14     protected void session(SelectionKey key) throws Exception {
15         SocketChannel channel = (SocketChannel) key.channel();
16         Communicator comm = (Communicator)key.attachment();
17         if (key.isReadable()) {
18             //接收文件名
```

```
19                if (comm.getStepIndex() == 0) {
20                    Receiver receiver = (Receiver) comm;
21                    if (receiver.recvFully(channel)){
22                        String fileName = (String)receiver.getObject();
23                        System.out.println("接收到文件名：" + fileName);
24                        String filePath = "D:\\ServerFiles\\FromClient-" + fileName;
25                        //下一次的操作还是 READ,只需修改 Attachment
26                        key.attach(new FileReceiver(receiver.getStepIndex() + 1, filePath));
27                    }
28                }
29                //接收文件
30                if (comm.getStepIndex() == 1) {
31                    FileReceiver receiver = (FileReceiver) comm;
32                    if (receiver.recvFully(channel)) {
33                        String s = receiver.getFilePath();
34                        System.out.println("接收文件完毕,保存至 " + s);
35                        //下一次的操作是 WRITE,需要重新注册
36                        channel.register(key.selector(),
                                  SelectionKey.OP_WRITE,
                                  new Sender(receiver.getStepIndex() + 1, "xyz.mp4"));
37                    }
38                }
39            }else{
40                //发送文件名
41                if (comm.getStepIndex() == 2){
42                    Sender sender = (Sender) comm;
43                    if (sender.sendFully(channel)){
44                        System.out.println("发送文件名：" + sender.getObject());
45                        String filePath = "D:\\ServerFiles\\" + sender.getObject();
46                        //下一次的操作还是 WRITE,不需要重复注册
47                        key.attach(new FileSender(sender.getStepIndex() + 1, filePath));
48                    }
49                }
50                //发送文件
51                if (comm.getStepIndex() == 3){
52                    FileSender sender = (FileSender) key.attachment();
53                    if (sender.sendFully(channel)) {
54                        System.out.println("发送文件成功：" + sender.getFilePath());
55                        this.closeChannel(key);
56                    }
57                }
58            }
59        }
60 }
```

Server3 对 Server 的 session(.)和 registerOnConnected(.)进行了重载。因为会话要求当双方建立连接后客户端发送而服务端接收，所以第 11 行 register(.)的第 2 个参数是 SelectionKey.OP_READ，第 3 个参数是 Receiver 对象，用于执行接收文件名的操作。

客户端代码见代码 2-14，对类 Client 进行扩展。

代码 2-14：类 Client3

```
1  public class Client3 extends Client{
2      public static void main(String[] args) throws Exception{
3          ExecutorService fixPool = Executors.newCachedThreadPool();
4          //为了避免单机上文件访问冲突，只用一个客户端
5          for (int i = 0;i < 1;i++) {
6              fixPool.execute(
7                      () ->{
8                          try {
9                              Client3 client = new Client3();
10                             client.communicate();
11                         }catch (Exception ex){
12                             ex.printStackTrace();
13                         }
14                     }
15             );
16         }
17         fixPool.shutdown();
18     }
19     public Client3() throws Exception{
20         super("127.0.0.1",8001);
21     }
22     @Override
23     protected void registerOnConnected(SocketChannel channel, Selector selector)
       throws Exception {
24         channel.register(selector, SelectionKey.OP_WRITE, new Sender(0,"abc.mp4"));
25     }
26     @Override
27     protected void session(SelectionKey key) throws Exception {
28         SocketChannel channel = (SocketChannel) key.channel();
29         Communicator comm = (Communicator)key.attachment();
30         if (key.isReadable()) {
31             //接收文件名
32             if (comm.getStepIndex() == 2) {
33                 Receiver receiver = (Receiver) comm;
34                 if (receiver.recvFully(channel)){
35                     String fileName = (String) receiver.getObject();
36                     System.out.println("接收到文件名：" + fileName);
37                     String filePath = "D:\\ClientFiles\\FromServer-" + fileName;
```

```
38                        //下一次的操作还是 READ,因此不用重复注册
39                        key.attach(new FileReceiver(receiver.getStepIndex() + 1, filePath));
40                    }
41                }
42                //接收文件
43                if (comm.getStepIndex() == 3) {
44                    FileReceiver receiver = (FileReceiver) comm;
45                    if (receiver.recvFully(channel)) {
46                        String s = receiver.getFilePath();
47                        System.out.println("接收文件完毕,保存至 " + s);
48                        this.closeChannel(key);
49                    }
50                }
51            }else{
52                //发送文件名
53                if (comm.getStepIndex() == 0){
54                    Sender sender = (Sender) comm;
55                    if (sender.sendFully(channel)){
56                        System.out.println("发送文件名: " + sender.getObject());
57                        String filePath = "D:\\ClientFiles\\" + sender.getObject();
58                        //不用重复注册
59                        key.attach(new FileSender(sender.getStepIndex() + 1, filePath));
60                    }
61                }
62                //发送文件
63                if (comm.getStepIndex() == 1){
64                    FileSender sender = (FileSender) key.attachment();
65                    if (sender.sendFully(channel)) {
66                        System.out.println("发送文件成功: " + sender.getFilePath());
67                        channel.register(key.selector(),
                                  SelectionKey.OP_READ,
                                  new Receiver(sender.getStepIndex() + 1, String.class));
68                    }
69                }
70            }
71        }
72 }
```

Client3 对 Client 的 session(.)和 registerOnConnected(.)进行了重载。因为会话要求当双方建立连接后客户端发送而服务端接收,所以第 24 行 register(.)的第 2 个参数是 SelectionKey.OP_WRITE,第 3 个参数是 Sender 对象,用于执行发送文件名的操作。

运行时先启动附件代码 nio 项目的 Server3,再启动附件代码 nio 项目的 Client3,服务端及客户端控制台分别显示如下信息。

服务端控制台显示的信息	客户端控制台显示的信息
服务器开启服务端口 8001,等待连接请求…… 接受客户端的连接,开始会话…… 接收到文件名:abc. mp4 接收文件完毕,保存至 D:\ServerFiles\FromClient-abc. mp4 发送文件名:xyz. mp4 发送文件成功:D:\ServerFiles\xyz. mp4 会话结束,服务端关闭连接	客户端连接成功,开始会话…… 发送文件名:abc. mp4 发送文件成功:D:\ClientFiles\abc. mp4 接收到文件名:xyz. mp4 接收文件完毕,保存至 D:\ClientFiles\FromServer-xyz. mp4 会话结束,客户端关闭连接

运行前要保证存在文件 D:\ClientFiles\abc. mp4 和 D:\ServerFiles\xyz. mp4。

2.9 设计多线程服务器

NIO 服务端的基本模式是单线程模式,即一个线程处理连接通道及所有的传输通道。虽然数据通道的处理是非阻塞的,但是其处理必然会影响连接通道的响应速度。为了提高服务端的响应速度,可以把连接通道与数据通道由不同的选择器来管理。一条基本的原则是选择器只能由一个线程遍历,以避免在 select(.)后遍历 SelectionKey 集合时造成冲突。因此有以下增强模式可供选择。

1. 连接通道增强

因为一个端口只能绑定一个连接通道,所以不论何种连接增强模式,连接通道只能有一个。连接通道的增强模式有两种,见图 2-9。

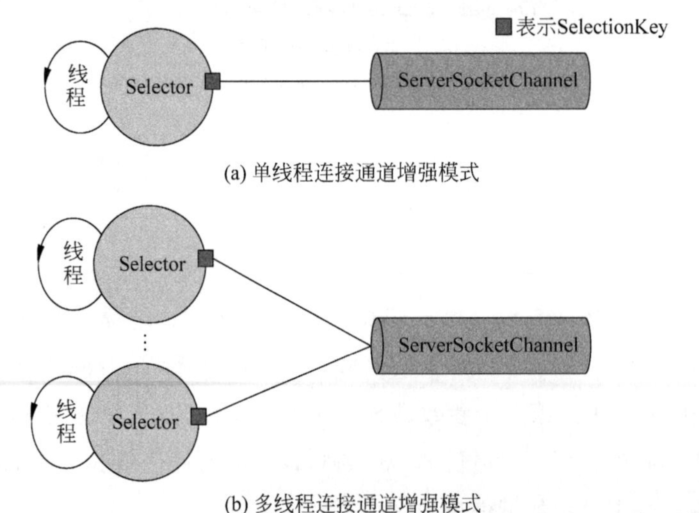

图 2-9 连接通道的两种增强模式

1）单线程连接通道增强模式

一个选择器只接受这个唯一的连接通道的注册，一个线程只处理该选择器通知的"接受连接就绪"事件，即只负责建立数据通道的连接。

2）多线程连接通道增强模式

这个唯一的连接通道向多个选择器注册，每个选择器由一个线程来处理该选择器通知的"接受连接就绪"事件，同样，这里的选择器只接受该连接通道的注册。

对于多线程连接通道增强模式，当客户端发出一个连接请求，在 ServerSocketChannel 上将出现"接受连接就绪"事件，这时，若恰逢多个选择器同时在执行 select(.)函数，则这些选择器都得到此事件通知，这将导致多个线程并发执行此连接通道的 accept(.)方法，这时只有一个线程会成功，获得一个新的数据通道，而其他不成功的线程获得 null 返回值。服务端的这种并发不会对客户端造成额外的影响，例如，服务端的并发线程之一已经 accept(.)成功，另一个线程再执行 accept(.)，不会影响刚才建立的数据通道连接。

2. 数据通道增强

在连接通道由另外的线程专门处理后，数据通道的增强模式有四种。

(1) 启动一个线程，该线程遍历一个选择器，而这个选择器接受所有数据通道的注册。

(2) 启动多个线程，每个线程遍历一个选择器，而这个选择器只接受一个数据通道的注册。

(3) 启动多个线程，每个线程遍历一个选择器，而这个选择器接受多个数据通道的注册。

(4) 启动多个线程，每个线程遍历一组选择器，而组内的每个选择器接受一个或多个数据通道的注册。

从图 2-10 可以直观地看出这 4 种模式的差异。

其中，实用的是模式 a 与模式 c，而模式 c 是模式 a 的扩展，更适用于高并发环境。本书提供数据通道增强模式 c 的代码实现以及连接通道的两种增强模式代码实现，同时图 2-5 所示的会话将作为运用实例。

图 2-11 是多线程服务相关的类图。类 Server4Worker 实现了数据通道增强模式 c，类 Server4Single 实现了连接通道增强模式 a，类 Server4Multiple 实现了连接通道增强模式 b。

类 Server4Worker 具有 Runnable 接口，显然每个 Server4Worker 实例将作为线程独立运行。Server4Worker 自带的私有成员变量 selector 作为其监听对象，其 run()方法将负责所有在此 selector 上注册的数据通道的 I/O 操作。服务端将启动多个 Server4Worker 实例线程，新建立的数据通道将被分派给某个 Server4Worker 线程。对于 Server4Worker 所管

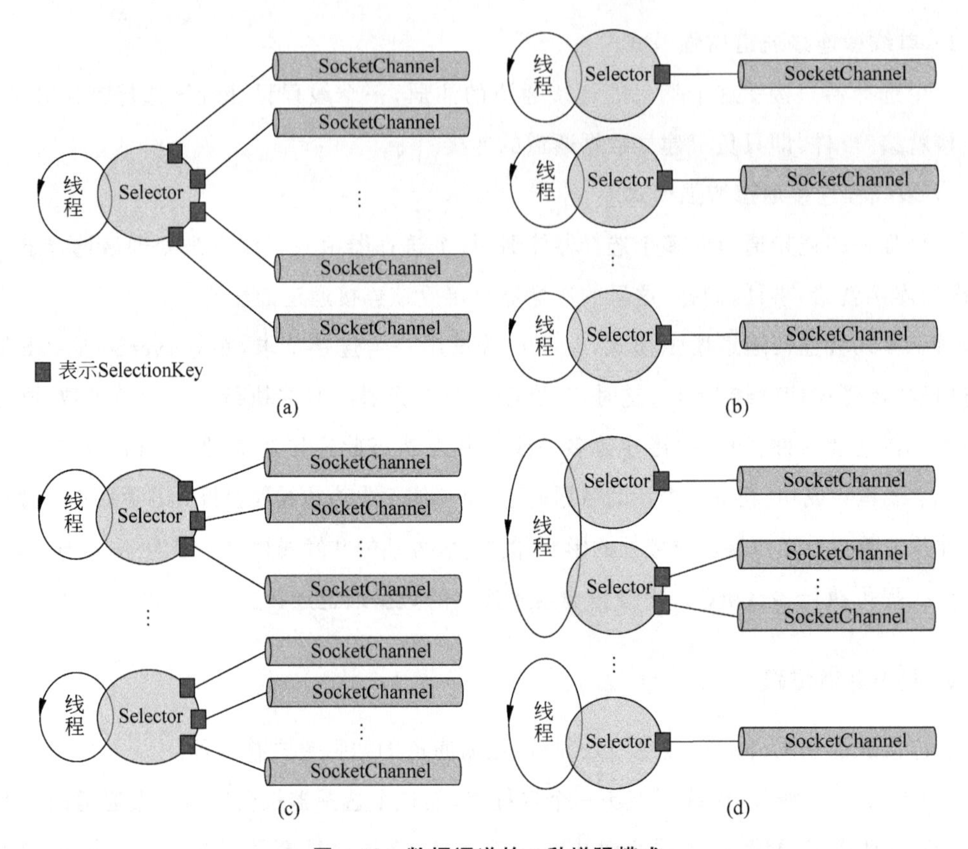

图 2-10　数据通道的 4 种增强模式

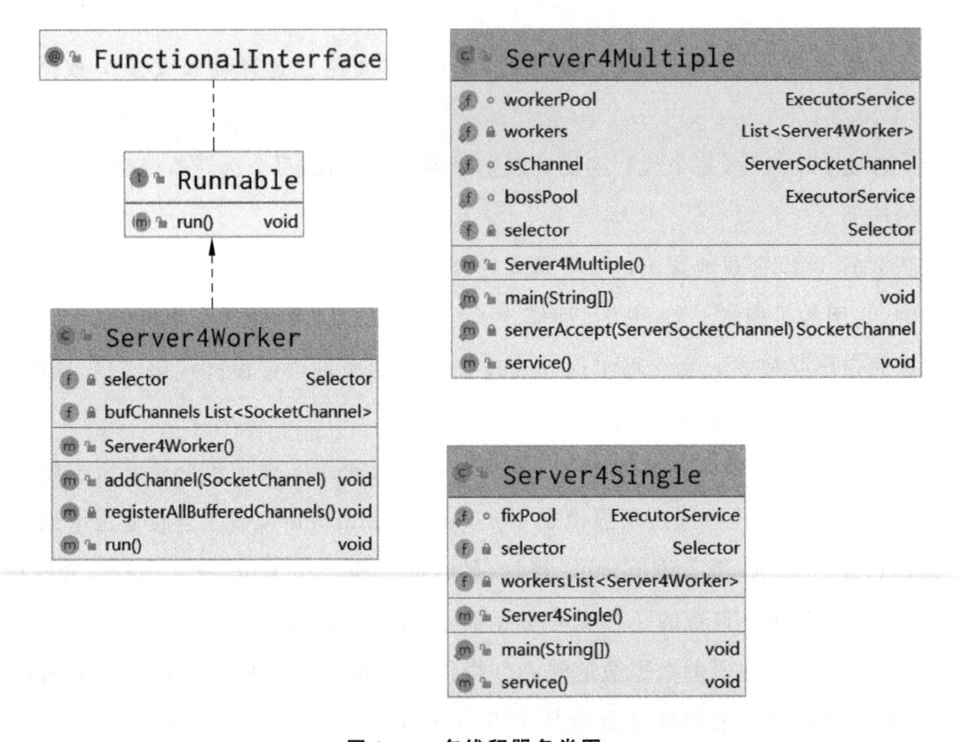

图 2-11　多线程服务类图

理的选择器来说，一方面要接受 Server4Worker 自身线程的 select(.)检查，另一方面要接受新派发的数据通道的注册，为了解决派发线程与 Server4Worker 线程的并发冲突问题，定义一个 List < SocketChannel >类型的成员变量 bufChannels，用于暂存派发给此 Server4Worker 线程的数据通道，并且在此 Server4Worker 线程每次 select(.)检查之前把暂存的多个数据通道向其管理的选择器注册。Server4Worker 的实现见代码 2-15。

代码 2-15：类 Server4Worker

```
1  public class Server4Worker implements Runnable {
2      //一个 Selector 管理多个 Channel
3      private Selector selector;
4      //用于暂存将要加入 selector 的数据通道
5      private List<SocketChannel> bufChannels = new ArrayList<>();
6      public Server4Worker() throws Exception{
7          selector = Selector.open();
8      }
9      synchronized public void addChannel(SocketChannel channel){
10         bufChannels.add(channel);
11     }
12     synchronized private void registerAllBufferedChannels() throws Exception{
13         Iterator<SocketChannel> iterator = bufChannels.iterator();
14         while (iterator.hasNext()) {
15             SocketChannel channel = iterator.next();
16             channel.configureBlocking(false);
17             channel.register(selector,
                               SelectionKey.OP_READ,
                               new Receiver(0,Object.class));
18             System.out.println("数据通道注册成功!");
19             iterator.remove();
20         }
21     }
22     private void closeChannel(SelectionKey key) throws Exception{
23         key.channel().close();
24         key.cancel();
25         System.out.println("会话结束,服务端关闭连接。");
26     }
27     @Override
28     public void run() {
29         while(true) {
30             try {
31                 registerAllBufferedChannels();
32                 selector.select(500);
33                 Set<SelectionKey> selectedKeysSet = selector.selectedKeys();
34                 Iterator<SelectionKey> iterator = selectedKeysSet.iterator();
```

```
35                      while (iterator.hasNext()) {
36                          SelectionKey key = iterator.next();
37                          iterator.remove();
38                          SocketChannel channel = (SocketChannel) key.channel();
39                          if (key.isReadable()) {
40                              Receiver receiver = (Receiver)key.attachment();
41                              if (receiver.recvFully(channel)){
42                                  Object obj = receiver.getObject();
43                                  System.out.println("接收对象：" + obj);
44                                  if (receiver.getStepIndex() == 0) {
45                                      channel.register(selector,
                                                SelectionKey.OP_WRITE,
                                                new Sender(
                                                        receiver.getStepIndex() + 1,
                                                        StuRepository.getStudent()));
46                                  }
47                                  if (receiver.getStepIndex() == 3){
48                                      closeChannel(key);
49                                  }
50                              }
51                          } else{
52                              Sender sender = (Sender)key.attachment();
53                              if (sender.sendFully(channel)){
54                                  System.out.println("发送对象：" + sender.getObject());
55                                  //根据步骤号设定 Sender 的信息
56                                  if (sender.getStepIndex() == 1) {
57                                      //因为允许的操作相同,只需修改 attachment
58                                      key.attach(new Sender(
                                                        sender.getStepIndex() + 1,
                                                        StuRepository.getStudent()));
59                                  }
60                                  if (sender.getStepIndex() == 2) {
61                                      channel.register(selector,
                                                SelectionKey.OP_READ,
                                                new Receiver(sender.getStepIndex() + 1,
                                                        Object.class));
62                                  }
63                              }
64                          }
65                      }
66                  } catch (Exception e) {
67                      e.printStackTrace();
68                  }
69              }
70          }
71  }
```

代码 2-15 第 32 行的超时参数很重要。如果不设置超时参数，则 select()将无限等待 I/O 事件的通知，而在阻塞时选择器检测不到新注册的数据通道上的事件，因此这个超时参数使得在长时间无连接的情况下出现新连接时依然可以正常通信。

代码 2-16 实现了连接通道增强模式 a，即连接通道只向唯一的选择器注册，因此只有一个线程负责处理所有的连接请求。

代码 2-16：类 Server4Single

```
1  public class Server4Single {
2      static final int port = 8001;
3      static ExecutorService fixPool = Executors.newCachedThreadPool();
4      public static void main(String[] args) throws Exception{
5          Server4Single server = new Server4Single();
6          server.service();
7      }
8      private Selector selector;
9      //一组处理数据通道 I/O 的线程，每个线程包含一个选择器
10     private List<Server4Worker> workers = new ArrayList<>();
11     public Server4Single() throws Exception{
12         //相当于启用 10 个选择器，来管理数据通道
13         for (int i = 0;i < 10;i++){
14             Server4Worker t = new Server4Worker();
15             workers.add(t);
16             fixPool.execute(t);
17         }
18         selector = Selector.open();
19         ServerSocketChannel ssChannel = ServerSocketChannel.open();
20         ssChannel.configureBlocking(false);
21         ssChannel.bind(new InetSocketAddress("127.0.0.1", port));
22         ssChannel.register(selector, SelectionKey.OP_ACCEPT);
23         System.out.println("服务端开启服务端口" + port + "……");
24     }
25     public void service() {
26         try {
27             while (true) {
28                 selector.select();
29                 Set<SelectionKey> selectedKeysSet = selector.selectedKeys();
30                 Iterator<SelectionKey> iterator = selectedKeysSet.iterator();
31                 while (iterator.hasNext()) {
32                     SelectionKey key = iterator.next();
33                     iterator.remove();
34                     if (key.isAcceptable()) {
35                         ServerSocketChannel ssChannel =
                                                (ServerSocketChannel) key.channel();
```

```
36                     SocketChannel channel = ssChannel.accept();
37                     System.out.println("接受客户端的连接,开始会话……");
38                     //随机选择一个 Worker 让它去处理 channel 的通信
39                     workers.get(new Random().nextInt(workers.size()))
                           .addChannel(channel);
40                 }
41             }
42         }
43     }catch (Exception ex){
44         ex.printStackTrace();
45     }
46   }
47 }
```

代码 2-16 的第 39 行是把新创建的数据通道添加到被随机选中的 Server4Worker 线程的通道暂存区。

代码 2-17 实现了连接通道增强模式 b,即连接通道向多个选择器注册,因此有多个线程负责处理连接请求。

代码 2-17:类 Server4Multiple

```
1  public class Server4Multiple {
2      static final int port = 8001;
3      static ExecutorService workerPool = Executors.newCachedThreadPool();
4      //一组处理数据通道 I/O 的线程,每个线程包含一个选择器
5      private static List<Server4Worker> workers = new ArrayList<>();
6      //绑定服务端口的 ServerChannel 只能有一个
7      static ServerSocketChannel ssChannel;
8      static ExecutorService bossPool = Executors.newCachedThreadPool();
9      public static void main(String[] args) throws Exception{
10         //相当于启用 50 个选择器,来管理数据通道
11         for (int i = 0;i < 50;i++){
12             Server4Worker t = new Server4Worker();
13             workers.add(t);
14             workerPool.execute(t);
15         }
16         //启用一组线程(5 个)来负责处理连接请求
17         for (int i = 0;i < 5;i++) {
18             bossPool.execute(
19                 () ->{
20                     try {
21                         Server4Multiple server = new Server4Multiple();
22                         server.service();
23                     }catch (Exception ex){
24                         ex.printStackTrace();
```

```
25                          }
26                      }
27              );
28          }
29          bossPool.shutdown();
30      }
31      static {
32          try {
33              //唯一的 ServerSocketChannel 初始化,绑定 IP 和端口
34              ssChannel = ServerSocketChannel.open();
35              ssChannel.configureBlocking(false);
36              ssChannel.bind(new InetSocketAddress("127.0.0.1", port));
37              System.out.println("服务端开启服务端口" + port + "……");
38          }catch (Exception e) {
39              e.printStackTrace();
40          }
41      }
42      private Selector selector;
43      //一个绑定了 IP 和端口的 ServerSocketChannel 向多个 Selector 注册
44      //而每个 Selector 由一个线程负责轮询
45      public Server4Multiple() throws Exception{
46          selector = Selector.open();
47          //唯一的 ServerSocketChannel 向每个 Boss 的 Selector 注册
48          ssChannel.register(selector, SelectionKey.OP_ACCEPT);
49      }
50      //此处无须 synchronized
51      private static SocketChannel serverAccept(ServerSocketChannel sChannel){
52          try {
53              SocketChannel channel = sChannel.accept();
54              return channel;
55          }catch (Exception e){
56              e.printStackTrace();
57              return null;
58          }
59      }
60      public void service() {
61          try {
62              while (true) {
63                  selector.select();
64                  Set<SelectionKey> selectedKeysSet = selector.selectedKeys();
65                  Iterator<SelectionKey> iterator = selectedKeysSet.iterator();
66                  while (iterator.hasNext()) {
67                      SelectionKey key = iterator.next();
68                      iterator.remove();
69                      if (key.isAcceptable()) {
```

```
70                         ServerSocketChannel ssChannel =
                                        (ServerSocketChannel) key.channel();
71                         SocketChannel channel = serverAccept(ssChannel);
72                         if (channel!= null) {
73                             System.out.println("接受客户端的连接,开始会话……");
74                             //随机选择一个 Worker 线程去处理 channel 的通信
75                             workers.get(new Random().nextInt(workers.size()))
                                    .addChannel(channel);
76                         }
77                     }
78                 }
79             }
80         }catch (Exception ex){
81             ex.printStackTrace();
82         }
83     }
84 }
```

代码 2-17 的第 75 行是把新创建的数据通道添加到被随机选中的 Server4Worker 线程的通道暂存区。

增强模式限于服务端,客户端不受影响,故客户端代码与代码 2-10 相同,这里不再赘述。

按照以下步骤分别测试连接通道的两种增强模式。

模式 a: 运行时先启动附件代码 nio 项目的 Server4Single,再启动附件代码 nio 项目的 Client4。

模式 b: 运行时先启动附件代码 nio 项目的 Server4Multiple,再启动附件代码 nio 项目的 Client4。

之后在服务端控制台与客户端控制台可以查看运行结果。

习题

编写 NIO 程序,完成如下功能: 客户端连接远程服务端,之后由使用者发送命令对服务端的文件系统进行操作,命令的顺序并不固定。

dir-列出当前文件夹内的文件及文件夹;

cd <文件夹>-改变服务端的当前文件夹;

download <远端文件名>-下载服务端的文件至本地;

upload <本地文件名>-上传本地文件至服务端;

close-结束会话。

第 3 章 AIO

AIO(asynchronous I/O)称为异步 I/O,采用异步回调方式实现非阻塞。AIO 与 NIO 的区别在于 NIO 选择器所发的通知是 I/O“就绪”事件,而 AIO 异步回调所发的通知是 I/O“完成”事件。AIO 可以用在请求并发量很高的网络环境。本章对应的源代码为附件中的 aio 项目。

3.1 异步操作概述

AIO 网络编程主要使用 AsynchronousServerSocketchannel 类和 AsychronousSocketchannel 类,前者是连接通道,后者是数据通道。客户端与服务端的通信过程与 BIO 及 NIO 相同,如图 3-1 所示。

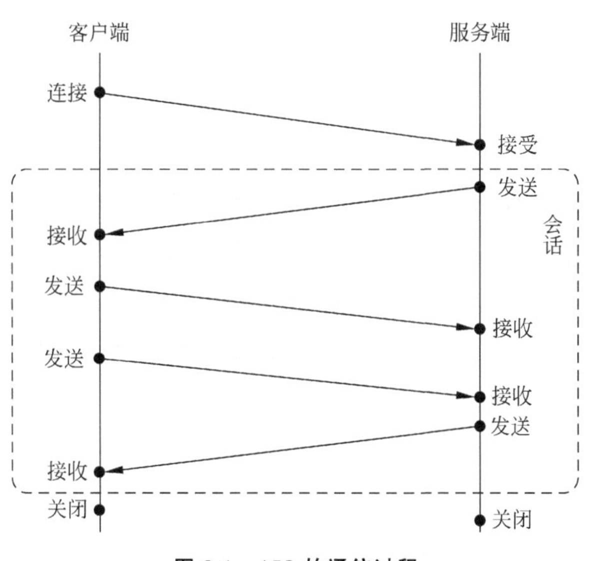

图 3-1 AIO 的通信过程

AIO 的特点是 accept(.)、connect(.)、read(.)、write(.)均为异步操作,这些函数的调用会立即返回,不会阻塞调用者线程;这些函数的真正执行是由 JVM 默认线程池的某个线程在后台完成的，而且，当操作完成之后，后台线程将自动执行由开发者预先提供的CompletionHandler接口实现。

```
public interface CompletionHandler<V,A> {
    void completed(V result, A attach);
    void failed(Throwable exc, A attach);
}
```

下面对上述4个异步函数分别说明。

1. accept(.)方法

```
public abstract <A> void accept(A attachment,
                    CompletionHandler<AsynchronousSocketChannel,? super A> handler);
```

当accept(.)在后台成功执行后，后台线程将自动执行参数handler的completed(.)方法，此时handler.completed(.)的result参数值被自动设为accept(.)成功执行所返回的AsynchronousSocketChannel即数据通道对象，handler.completed(.)的attach参数值被自动设为调用accept(.)时的attachment参数值；如果accept(.)在后台执行时抛出异常，则后台程序会自动执行handler的failed(.)方法，此时handler.failed(.)的exc参数值被自动设为所抛出的异常对象，handler.failed(.)的attach参数值被自动设为调用accept(.)时的attachment参数值。

2. connect(.)方法

```
public abstract <A> void connect(SocketAddress remote,A attachment,
                                 CompletionHandler<Void,? super A> handler);
```

当connect(.)在后台成功执行后，后台线程将自动执行参数handler的completed(.)方法，此时handler.completed(.)的result参数值被自动设为null，因为connect(.)成功执行返回的是null，handler.completed(.)的attach参数值被自动设为调用connect(.)时的attachment参数值；如果connect(.)在后台执行时抛出异常，则后台程序会自动执行handler的failed(.)方法，此时handler.failed(.)的exc参数值被自动设为所抛出的异常对象，handler.failed(.)的attach参数值被自动设为调用connect(.)时的attachment参数值。

3. read(.)方法

```
public final <A> void read(ByteBuffer dst,A attachment,
                           CompletionHandler<Integer,? super A> handler);
```

调用read(.)时，参数dst用来存放将要接收到的字节。当read(.)在后台成功执行后，所接收到的字节被存放于参数 dst,并且后台线程将自动执行参数 handler 的 completed(.)方法,此时 handler. completed(.)的 result 参数值被自动设为 read(.)成功执行所返回的接收字节数,handler. completed(.)的 attach 参数值被自动设为调用 read(.)时的 attachment 参数值；如果 read(.)在后台执行时抛出异常,则后台程序会自动执行 handler 的 failed(.)方法,此时 handler. failed(.)的 exc 参数值被自动设为所抛出的异常对象,handler. failed(.)的 attach 参数值被自动设为调用 read(.)时的 attachment 参数值。

虽然是异步操作,但是 read(.)并不能保证把参数 dst 的可用空间读满。

4. write(.)方法

```
public final <A> void write(ByteBuffer src,A attachment,
                            CompletionHandler<Integer,? super A> handler)
```

调用 write(.)时,参数 src 用来存放将要发送的字节。当 write(.)在后台成功执行后,存放于 src 的字节被发送出去,并且后台线程将自动执行参数 handler 的 completed(.)方法,此时 handler. completed(.)的 result 参数值被自动设为 write(.)成功执行所返回的发送字节数,handler. completed(.)的 attach 参数值被自动设为调用 write(.)时的 attachment 参数值；如果 write(.)在后台执行时抛出异常,则后台程序会自动执行 handler 的 failed(.)方法,此时 handler. failed(.)的 exc 参数值被自动设为所抛出的异常对象,handler. failed(.)的 attach 参数值被自动设为调用 write(.)时的 attachment 参数值。

虽然是异步操作,但是 write(.)并不能保证把参数 src 的可用字节全部发送出去。

3.2 AIO 服务端框架代码

服务端的操作包括开启侦听端口,接收来自客户端的连接请求,然后进行会话,会话完毕时关闭数据通道,见代码 3-1。

代码 3-1：类 Server

```
1  public abstract class Server {
2      private AsynchronousServerSocketChannel serverSocketChannel;
3      protected Server(int port) throws Exception{
4          serverSocketChannel = AsynchronousServerSocketChannel.open()
5                  .bind(new InetSocketAddress(port));
6          System.out.println("服务器开启服务端口,等待连接请求……");
7      }
8      public void service() {
9          //accept()交给线程池执行
10         serverSocketChannel.accept(null,
```

```
11                   new CompletionHandler < AsynchronousSocketChannel, Void >() {
12                       @Override
13                       public void completed(
                                    AsynchronousSocketChannel channel,
                                    Void att) {
14                           serverSocketChannel.accept(null, this);
15                           try{
16                               System.out.println(
                                        "服务器接受连接请求,开始会话……");
17                               session(channel);
18                               channel.close();
19                               System.out.println("会话结束,服务器关闭连接!");
20                           }catch (Exception e){
21                               e.printStackTrace();
22                           }
23                       }
24                       @Override
25                       public void failed(Throwable exc, Void att) {
26                           exc.printStackTrace();
27                       }
28                   });
29           try {                               //防止主线程退出
30               Thread.currentThread().join();
31           } catch (InterruptedException e) { }
32       }
33       protected abstract void session(
                     AsynchronousSocketChannel channel) throws Exception;
34 }
```

Server 类使用了"模板方法"设计模式,表示会话的抽象方法 session(.)将在子类中得到实现。不需要为了提高服务端响应速度而显式地使用线程池,因为异步调用 accept(.)已经隐式地使用了 JVM 默认线程池;也不需要把 accept(.)放在 while(true)循环里,因为异步式递归调用代替了 while 循环,见代码第 10 行和第 14 行。对于使用回调方法的异步函数 accept(.),不是由主线程取得函数的返回值,而是在函数结束时的回调方法里直接使用该函数的返回值。

3.3 AIO 客户端框架代码

客户端的操作包括连接服务端,然后进行会话,会话完毕时关闭数据通道,见代码 3-2。

代码 3-2:类 Client

```
1  public abstract class Client {
2      private String serverIP;
```

```
3       private int port;
4       protected Client(String serverIP,int port){
5           this.serverIP = serverIP;
6           this.port = port;
7       }
8       public void communicate() throws Exception{
9           CountDownLatch end = new CountDownLatch(1);
10          AsynchronousSocketChannel socketChannel =
                                    AsynchronousSocketChannel.open();
11          //connect(.)交给线程池执行
12          socketChannel.connect(new InetSocketAddress(serverIP, port), null,
                    new CompletionHandler<Void, Void>() {
14                      @Override
15                      public void completed(Void result, Void att) {
16                          try {
17                              System.out.println("客户端连接成功,开始会话……");
18                              session(socketChannel);
19                              socketChannel.close();
20                              System.out.println("会话结束,客户端关闭连接!");
21                          }catch (Exception e){
22                              e.printStackTrace();
23                          }finally {
24                              end.countDown();
25                          }
26                      }
27                      @Override
28                      public void failed(Throwable exc, Void att) {
29                          exc.printStackTrace();
30                          end.countDown();
31                      }
32                  });
33          //等待,直至连接关闭,然后结束程序
34          end.await();
35      }
36      protected abstract void session(
                AsynchronousSocketChannel channel) throws Exception;
37 }
```

Client 类使用了“模板方法”设计模式,表示会话的抽象方法 session(.)将在子类中得到实现。

3.4 AIO 的分帧问题

AIO 依然需要解决分帧问题,即接收方需要知道接收到的字节序列中正确的帧分界点。为此定义函数 recvBufferFully(.),见代码 3-3。

代码 3-3：函数 recvBufferFully(.)

```
1  private static ByteBuffer recvBufferFully(AsynchronousSocketChannel channel)
           throws Exception{
2      Attachment att = new Attachment(new CountDownLatch(1), null);
3      ByteBuffer countBuffer = ByteBuffer.allocate(4);
4      //接收表示长度的 4 字节
5      channel.read(countBuffer, countBuffer,
           new CompletionHandler<Integer, ByteBuffer>() {
6          @Override
7          public void completed(Integer result, ByteBuffer countBuffer) {
8              //确保读满 4 字节
9              if (countBuffer.hasRemaining()){
10                 channel.read(countBuffer, countBuffer, this);
11                 return;
12             }
13             att.setBuffer(ByteBuffer.allocate(countBuffer.getInt(0)));
14             channel.read(att.getBuffer(), att,
                   new CompletionHandler<Integer, Attachment>() {
15                 @Override
16                 public void completed(Integer result, Attachment att) {
17                     //确保读满数据缓冲区
18                     if (att.getBuffer().hasRemaining()){
19                         channel.read(att.getBuffer(), att, this);
20                         return;
21                     }
22                     att.getEndLatch().countDown();
23                 }
24                 @Override
25                 public void failed(Throwable exc, Attachment att) {
26                     exc.printStackTrace();
27                     att.getEndLatch().countDown();
28                 }
29             });
30         }
31         @Override
32         public void failed(Throwable exc, ByteBuffer attachment) {
33             exc.printStackTrace();
34             att.getEndLatch().countDown();
35         }
36     });
37     att.getEndLatch().await();
38     return att.getBuffer();
39 }
```

recvBufferFully(.)函数以异步的方式完成以下步骤。

(1) 接收一个整数。

(2) 构建一个缓冲区，大小为第(1)步的整数。

(3) 把缓冲区读满。

这个函数是接收对象、文件等操作的基础。

为了与接收方的 recvBufferFully(.)操作对应，代码 3-4 是为发送方定义的发送函数 sendBufferFully(.)。

代码 3-4：函数 sendBufferFully(.)

```
1  private static void sendBufferFully(
            AsynchronousSocketChannel channel, ByteBuffer buf) throws Exception{
2      Attachment att = new Attachment(new CountDownLatch(1), buf);
3      //countBuffer 用于存放要发送的字节数
4      ByteBuffer countBuffer = ByteBuffer.allocate(4);
5      countBuffer.putInt(buf.capacity());
6      countBuffer.flip();
7      //先发送表示缓冲区长度的 4 字节
8      channel.write(countBuffer, countBuffer,
           new CompletionHandler<Integer, ByteBuffer>() {
9          @Override
10         public void completed(Integer result, ByteBuffer countBuffer) {
11             if (countBuffer.hasRemaining()) {
12                 channel.write(countBuffer, countBuffer, this);
13                 return;
14             }
15             //表示长度的 4 字节已发送完毕，再发送缓冲区内容
16             channel.write(att.getBuffer(), att,
                   new CompletionHandler<Integer, Attachment>() {
17                 @Override
18                 public void completed(Integer result, Attachment att) {
19                     if (att.getBuffer().hasRemaining()) {
20                         channel.write(att.getBuffer(), att, this);
21                         return;
22                     }
23                     //缓冲区已发送完毕
24                     att.getEndLatch().countDown();
25                 }
26                 @Override
27                 public void failed(Throwable exc, Attachment att) {
28                     try{
29                         channel.close();
30                     }catch (Exception e){ }
```

```
31                      exc.printStackTrace();
32                      att.getEndLatch().countDown();
33                  }
34              });
35          }
36          @Override
37          public void failed(Throwable exc, ByteBuffer attachment) {
38              try{
39                  channel.close();
40              }catch (Exception e){ }
41              exc.printStackTrace();
42              att.getEndLatch().countDown();
43          }
44      });
45      //等待缓冲区发送完毕
46      att.getEndLatch().await();
47 }
```

sendBufferFully(.)函数使用回调的方式,完整地发送一个 Buffer。该函数是阻塞式的,当函数返回时,Buffer 已经整体发送;若这个函数处于阻塞状态,则还处于发送中。这种同步效果并不是通过轮询达到的,而是用异步方式和 CountDownLatch 实现的。这个函数是发送对象、文件等操作的基础。

上述两个函数使用了类 Attachment,该类封装了 ByteBuffer,并提供 CountDownLatch 用于控制收、发时的同步,见代码 3-5。

代码 3-5:类 Attachment

```
1  @Data
2  @AllArgsConstructor
3  @NoArgsConstructor
4  public class Attachment {
5      protected CountDownLatch endLatch;
6      protected ByteBuffer buffer;
7  }
```

3.5 案例 1:传输字符串的会话

本案例完成如图 3-2 所示的传输字符串的会话。

为发送方定义发送字符串的函数 sendString(.)见代码 3-6。该函数调用了代码 3-4 的 sendBufferFully(.)。

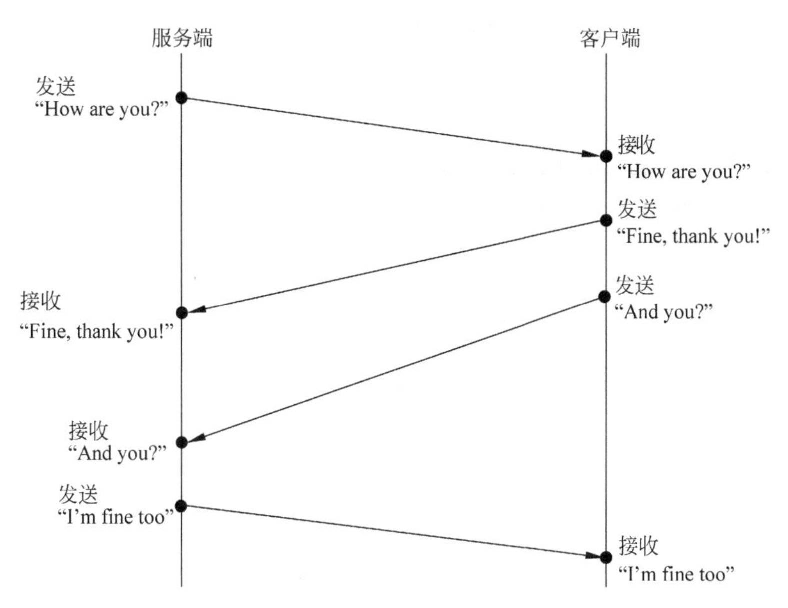

图 3-2 传输字符串的会话流程

代码 3-6：函数 sendString(.)

```
1  public static void sendString(AsynchronousSocketChannel channel, String msg)
                throws Exception {
2      ByteBuffer buffer = ByteBuffer.wrap(msg.getBytes("UTF-8"));
3      sendBufferFully(channel, buffer);
4  }
```

为接收方定义接收字符串的函数 recvString(.)见代码 3-7。该函数调用了代码 3-3 的 recvBufferFully(.)。

代码 3-7：函数 recvString(.)

```
1  public static String recvString(AsynchronousSocketChannel channel) throws Exception {
2      ByteBuffer buffer = recvBufferFully(channel);
3      buffer.flip();
4      byte[] bytes = buffer.array();
5      return new String(bytes,"UTF-8");
6  }
```

服务端 Server1 代码见代码 3-8。

代码 3-8：类 Server1

```
1  public class Server1 extends Server{
2      public static void main(String[] args) throws Exception {
3          Server1 server = new Server1();
4          server.service();
5      }
```

```
6       public Server1() throws Exception{
7           super(8001);
8       }
9       @Override
10      protected void session(AsynchronousSocketChannel channel) throws Exception {
11          String str = "How are you?";
12          Sender.sendString(channel, str);
13          System.out.println("成功发送：" + str);
14          str = Receiver.recvString(channel);
15          System.out.println("成功接收：" + str);
16          str = Receiver.recvString(channel);
17          System.out.println("成功接收：" + str);
18          str = "I'm fine too.";
19          Sender.sendString(channel, str);
20          System.out.println("成功发送：" + str);
21      }
22 }
```

Server1 重载了 Server 的 session(.)方法，该方法实现了服务端的会话流程。

客户端 Client1 代码见代码 3-9。

代码 3-9：类 Client1

```
1  public class Client1 extends Client{
2      public static void main(String[] args) {
3          ExecutorService fixPool = Executors.newCachedThreadPool();
4          for (int i = 0; i < 10; i++) {
5              fixPool.execute(
6                      () -> {
7                          try {
8                              Client1 client = new Client1();
9                              client.communicate();
10                         } catch (Exception ex) {
11                             ex.printStackTrace();
12                         }
13                     }
14             );
15         }
16         fixPool.shutdown();
17     }
18     public Client1(){
19         super("127.0.0.1", 8001);
20     }
21     @Override
22     protected void session(AsynchronousSocketChannel channel) throws Exception {
23         String str = Receiver.recvString(channel);
```

```
24         System.out.println("成功接收：" + str);
25         str = "I am fine. Thank you!";
26         Sender.sendString(channel, str);
27         System.out.println("成功发送：" + str);
28         str = "And you?";
29         Sender.sendString(channel, str);
30         System.out.println("成功发送：" + str);
31         str = Receiver.recvString(channel);
32         System.out.println("成功接收：" + str);
33     }
34 }
```

Client1 重载了 Client 的 session(.)方法，该方法实现了客户端的会话流程。

运行时，先启动附件代码 aio 项目的 Server1，再启动附件代码 aio 项目的 Client1。

3.6 案例 2：传输对象的会话

当客户端与服务端连接成功，它们之间进行如图 3-3 所示的会话。

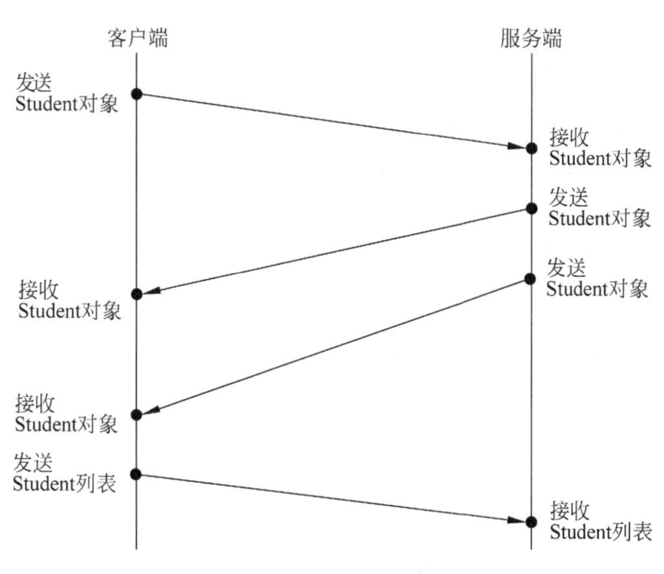

图 3-3 传输对象的会话流程

服务端：接收对象，发送对象，发送对象，接收对象列表。

客户端：发送对象，接收对象，接收对象，发送对象列表。

为发送方定义发送对象的函数 sendObjectSerializable(.)见代码 3-10，该函数调用了代码 3-4 的 sendBufferFully(.)及代码 1-14 的序列化函数 objSerializableToByteArray(.)。这里要求发送的对象具有 Serializable 接口。

代码 3-10：函数 sendObjectSerializable(.)

```
1  public static void sendObjSerializable(AsynchronousSocketChannel channel, Object obj)
           throws Exception{
2      ByteBuffer buf = ByteBuffer.wrap(Utils.objSerializableToByteArray(obj));
3      sendBufferFully(channel, buf);
4  }
```

为接收方定义接收对象的函数 recvObjectSerializable(.)见代码 3-11，该函数调用了代码 3-3 的 recvBufferFully(.)及代码 1-15 的反序列化函数 byteArrayToObjSerializable(.)。这里要求接收的对象具有 Serializable 接口。

代码 3-11：函数 recvObjectSerializable(.)

```
1  public static Object recvObjSerializable(AsynchronousSocketChannel channel)
           throws Exception{
2      ByteBuffer buf = recvBufferFully(channel);
3      buf.flip();
4      byte[] bytes = buf.array();
5      return Utils.byteArrayToObjSerializable(bytes);
6  }
```

服务端 Server2 代码见代码 3-12。

代码 3-12：类 Server2

```
1  public class Server2 extends Server{
2      public static void main(String[] args) throws Exception {
3          Server2 server = new Server2();
4          server.service();
5      }
6      public Server2() throws Exception{
7          super(8001);
8      }
9      @Override
10     protected void session(AsynchronousSocketChannel channel) throws Exception {
11         Student stu;
12         stu = (Student) Receiver.recvObjSerializable(channel);
13         System.out.println("接收到对象：" + stu);
14         stu = StuRepository.getStudent();
15         Sender.sendObjSerializable(channel, stu);
16         System.out.println("发送了对象：" + stu);
17         stu = StuRepository.getStudent();
18         Sender.sendObjSerializable(channel, stu);
19         System.out.println("发送了对象：" + stu);
20         List<Student> students =
                           (List<Student>)Receiver.recvObjSerializable(channel);
```

```
21          System.out.println("接收了列表：" + students);
22      }
23 }
```

Server2 重载了 Server 的 session(.)方法，该方法实现了服务端的会话流程。

客户端 Client2 代码见代码 3-13。

代码 3-13：类 Client2

```
1  public class Client2 extends Client{
2      public static void main(String[] args) {
3          ExecutorService fixPool = Executors.newCachedThreadPool();
4          for (int i = 0; i < 10; i++) {  //模拟 10 个客户端并发操作
5              fixPool.execute(
6                      () -> {
7                          try {
8                              Client2 client = new Client2();
9                              client.communicate();
10                         } catch (Exception ex) {
11                             ex.printStackTrace();
12                         }
13                     }
14             );
15         }
16         fixPool.shutdown();
17     }
18     public Client2(){
19         super("127.0.0.1", 8001);
20     }
21     @Override
22     protected void session(AsynchronousSocketChannel channel) throws Exception {
23         Student stu = StuRepository.getStudent();
24         Sender.sendObjSerializable(channel, stu);
25         System.out.println("发送了对象：" + stu);
26         stu = (Student) Receiver.recvObjSerializable(channel);
27         System.out.println("接收到对象：" + stu);
28         stu = (Student) Receiver.recvObjSerializable(channel);
29         System.out.println("接收到对象：" + stu);
30         List<Student> list = StuRepository.getStudents();
31         Sender.sendObjSerializable(channel, list);
32         System.out.println("发送了列表：" + list);
33     }
34 }
```

Client2 重载了 Client 的 session(.)方法，该方法实现了客户端的会话流程。

运行时，先启动附件代码 aio 项目的 Server2，再启动附件代码 aio 项目的 Client2。

3.7 案例 3：传输文件的会话

当客户端与服务端连接成功，它们之间进行如图 3-4 所示的会话。

服务端：接收文件名，接收文件(并保存)，发送文件名，发送文件内容。

客户端：发送文件名，发送文件内容，接收文件名，接收文件(并保存)。

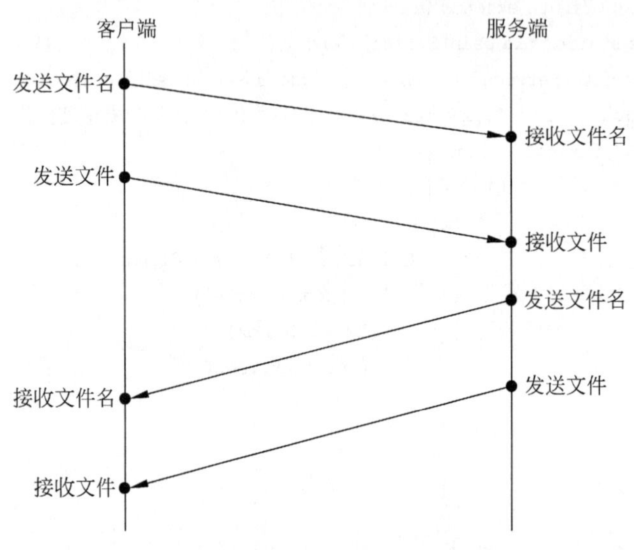

图 3-4 传输文件的会话流程

为了辅助文件的发送，定义 AttachmentForFile，见代码 3-14。它是代码 3-5 定义的 Attachment 的子类，增加了文件读取位置以及剩余未读字节数这两个信息。

代码 3-14：类 AttachmentForFile

```
1  @Data
2  public class AttachmentForFile extends Attachment {
3      private long posInFile;
4      private long restByteCount;
5      public AttachmentForFile(CountDownLatch endLatch,
                   ByteBuffer buffer, long posInFile, long restByteCount) {
6          this.endLatch = endLatch;
7          this.buffer = buffer;
8          this.posInFile = posInFile;
9          this.restByteCount = restByteCount;
10     }
11 }
```

对于大文件的传输，依然采用分片传输的方式，这与之前 BIO 或者 NIO 传输大文件的思路是一样的，有个不同的细节是，这里对文件也是采用 AIO 进行读写。传输大文件的处理流程如图 3-5 所示。

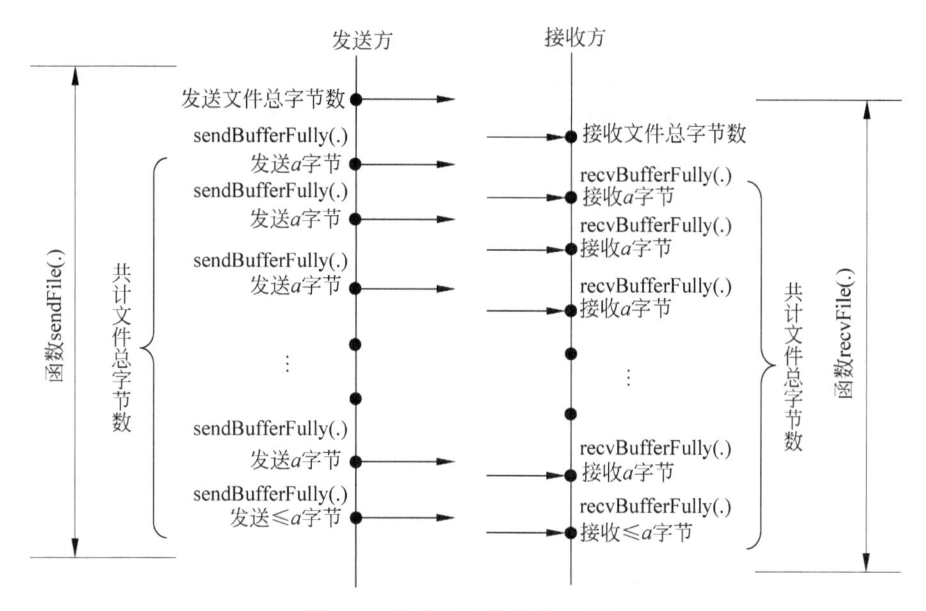

图 3-5　传输大文件的处理流程

函数 sendFile(.)的定义参见代码 3-15。sendFile(.)是阻塞式同步函数，只有当文件内容全部发送完毕，函数才返回，否则函数阻塞。

代码 3-15：函数 sendFile(.)

```
1     public static void sendFile(AsynchronousSocketChannel channel, String filePath)
      throws Exception {
2       Path path = Paths.get(filePath);
3       AsynchronousFileChannel fileChannel =
                    AsynchronousFileChannel.open(path, StandardOpenOption.READ);
4       long restByteCount = fileChannel.size();
5       //countBuffer 用于存放要发送的文件长度值
6       ByteBuffer countBuffer = ByteBuffer.allocate(8);
7       countBuffer.putLong(restByteCount);
8       countBuffer.flip();
9       CountDownLatch endHeader = new CountDownLatch(1);
10      //先发送总字节数
11      channel.write(countBuffer, countBuffer,
            new CompletionHandler<Integer, ByteBuffer>() {
12          @Override
13          public void completed(Integer result, ByteBuffer attachment) {
14              if (countBuffer.hasRemaining()) {
15                  channel.write(countBuffer, countBuffer, this);
16                  return;
17              }
18              endHeader.countDown();
```

```
19            }
20            @Override
21            public void failed(Throwable exc, ByteBuffer attachment) {
22                exc.printStackTrace();
23                endHeader.countDown(); //避免发生异常时死锁
24            }
25        });
26        //同步：保证先发送总字节数，再发送数据字节
27        //因为共用同一个 SocketChannel
28        endHeader.await();
29        //从文件读取数据再发送
30        int buffSize = 1024 * 1024;
31        if (buffSize > restByteCount) {
32            buffSize = (int) restByteCount;
33        }
34        AttachmentForFile att = new AttachmentForFile(new CountDownLatch(1),
                                      ByteBuffer.allocate(buffSize), 0, restByteCount);
35        fileChannel.read(att.getBuffer(), att.getPosInFile(), att,
              new CompletionHandler<Integer, AttachmentForFile>() {
36            @Override
37            public void completed(Integer result, AttachmentForFile att) {
38                att.setPosInFile(att.getPosInFile() + result);
39                att.setRestByteCount(att.getRestByteCount() - result);
40                //缓冲区未满，继续从文件填入
41                if (att.getBuffer().hasRemaining()) {
42                    fileChannel.read(att.getBuffer(), att.getPosInFile(), att, this);
43                    return;
44                }
45                //缓冲区已满，先写出至 socketChannel
46                att.getBuffer().flip();
47                try {
48                    sendBufferFully(channel, att.getBuffer());
49                } catch (Exception e) {
50                    e.printStackTrace();
51                }
52                //检查是否已经到达文件末尾，
53                //已经到达文件末尾，则当前函数 sendFileFully(.)结束
54                if (att.getRestByteCount() == 0) {
55                    att.getEndLatch().countDown();
56                    return;
57                }
58                //还未到达文件末尾，则清空缓冲区，再次填写缓冲区
59                att.getBuffer().clear();
60                if (att.getBuffer().capacity() > att.getRestByteCount()) {
61                    att.setBuffer(ByteBuffer.allocate((int) att.getRestByteCount()));
```

```
62                }
63                fileChannel.read(att.getBuffer(), att.getPosInFile(), att, this);
64            }
65            @Override
66            public void failed(Throwable exc, AttachmentForFile att) {
67                exc.printStackTrace();
68                att.getEndLatch().countDown(); //避免发生异常时死锁
69            }
70        });
71        //等待文件发送完成
72        att.getEndLatch().await();
73        fileChannel.close();
74 }
```

函数 recvFile(.)的定义参见代码 3-16。recvFile(.)是阻塞式同步函数，只有当文件内容全部接收完毕，函数才返回，否则函数阻塞。

代码 3-16：函数 recvFile(.)

```
1     public static void recvFile(AsynchronousSocketChannel channel, String filePath)
      throws Exception {
2      Path path = Paths.get(filePath);
3      AsynchronousFileChannel fileChannel = AsynchronousFileChannel.open(path,
                     StandardOpenOption.WRITE,  StandardOpenOption.CREATE);
4       //先接收一个长整数，表示总的字节数
5       ByteBuffer countBuffer = ByteBuffer.allocate(8);
6       CountDownLatch endHeader = new CountDownLatch(1);
7       channel.read(countBuffer, countBuffer,
            new CompletionHandler<Integer, ByteBuffer>() {
8           @Override
9           public void completed(Integer result, ByteBuffer attachment) {
10              //确保接收完 8 字节
11              if (countBuffer.hasRemaining()){
12                  channel.read(countBuffer, countBuffer, this);
13                  return;
14              }
15              endHeader.countDown();
16          }
17          @Override
18          public void failed(Throwable exc, ByteBuffer attachment) {
19              exc.printStackTrace();
20              endHeader.countDown(); //避免发生异常时死锁
21          }
22      });
23      //同步：必须等待总字节数接收完毕，再接收文件数据
24      //因为多线程共用一个 SocketChannel，而且有了总字节数才能做后续处理
```

```
25      endHeader.await();
26      ByteBuffer buffer = recvBufferFully(channel);
27      buffer.flip();
28      AttachmentForFile att = new AttachmentForFile(new CountDownLatch(1),
                                            buffer, 0L, countBuffer.getLong(0));
29      //异步方式保存文件
30      fileChannel.write(att.getBuffer(), att.getPosInFile(), att,
            new CompletionHandler<Integer, AttachmentForFile>() {
31          @Override
32          public void completed(Integer result, AttachmentForFile att) {
33              att.setPosInFile(att.getPosInFile() + result);
34              att.setRestByteCount(att.getRestByteCount() - result);
35              //确保缓冲区内容全部写入文件
36              if (att.getBuffer().hasRemaining()){
37                  fileChannel.write(att.getBuffer(), att.getPosInFile(), att, this);
38                  return;
39              }
40              //全部的字节写完了,结束函数 recvFileFully()
41              if (att.getRestByteCount() == 0){
42                  att.getEndLatch().countDown();
43                  return;
44              }
45              //尚未全部写完,则先读入,再次写入文件
46              ByteBuffer buffer = null;
47              try {
48                  buffer = recvBufferFully(channel);
49              }catch (Exception e){ e.printStackTrace();}
50              buffer.flip();
51              att.setBuffer(buffer);
52              fileChannel.write(att.getBuffer(), att.getPosInFile(), att, this);
53          }
54          @Override
55          public void failed(Throwable exc, AttachmentForFile att) {
56              att.getEndLatch().countDown(); //避免出现异常时发生死锁
57          }
58      });
59      att.getEndLatch().await();
60      fileChannel.close();
61 }
```

服务端 Server3 代码参见代码 3-17。

代码 3-17：类 Server3

```
1  public class Server3 extends Server{
2      public static void main(String[] args) throws Exception {
```

```
3            Server3 server = new Server3();
4            server.service();
5        }
6        public Server3() throws Exception{
7            super(8001);
8        }
9        @Override
10       protected void session(AsynchronousSocketChannel channel) throws Exception {
11           //底层是异步,效果是同步
12           String fileName = Receiver.recvString(channel);
13           System.out.println("接收文件名:" + fileName);
14           String filePath = "D:\\ServerFiles\\FromClient-" + fileName;
15           Receiver.recvFile(channel, filePath);
16           System.out.println("接收并保存文件:" + filePath);
17           filePath = "D:\\ServerFiles\\xyz.mp4";
18           fileName = filePath.substring(filePath.lastIndexOf("\\") + 1);
19           Sender.sendString(channel, fileName);
20           System.out.println("发送文件名:" + fileName);
21           Sender.sendFile(channel,filePath);
22           System.out.println("发送文件:" + filePath);
23       }
24 }
```

Server3 重载了 Server 的 session(.)方法,该方法实现了服务端的会话流程。

客户端 Client3 代码参见代码 3-18。

代码 3-18:类 Client3

```
1  public class Client3 extends Client{
2      public static void main(String[] args) {
3          ExecutorService fixPool = Executors.newCachedThreadPool();
4          for (int i = 0; i < 1; i++) {
5              fixPool.execute(
6                      () -> {
7                          try {
8                              Client3 client = new Client3();
9                              client.communicate();
10                         } catch (Exception ex) {
11                             ex.printStackTrace();
12                         }
13                     }
14             );
15         }
16         fixPool.shutdown();
17     }
18     public Client3(){
```

```
19          super("127.0.0.1", 8001);
20      }
21      @Override
22      protected void session(AsynchronousSocketChannel channel) throws Exception {
23          //底层是异步,效果是同步
24          String filePath = "D:\\ClientFiles\\abc.mp4";
25          String fileName = filePath.substring(filePath.lastIndexOf("\\") + 1);
26          Sender.sendString(channel, fileName);
27          System.out.println("发送文件名:" + fileName);
28          Sender.sendFile(channel,filePath);
29          System.out.println("发送文件:" + filePath);
30          fileName = Receiver.recvString(channel);
31          System.out.println("接收文件名:" + fileName);
32          filePath = "D:\\ClientFiles\\FromServer-" + fileName;
33          Receiver.recvFile(channel, filePath);
34          System.out.println("接收并保存文件:" + filePath);
35 }
```

Client3 重载了 Client 的 session(.)方法,该方法实现了客户端的会话流程。

运行时,先启动附件代码 aio 项目的 Server3,再启动附件代码 aio 项目的 Client3。

习题

1. 编写 AIO 程序,完成如下功能:客户端连接远程服务端,之后由使用者发送命令对服务端的文件系统进行操作,命令的顺序并不固定。

dir-列出当前文件夹内的文件及文件夹;

cd <文件夹>-改变服务端的当前文件夹;

download <远端文件名>-下载服务端的文件至本地;

upload <本地文件名>-上传本地文件至服务端;

close-结束会话。

2. 改写代码 3-15 的 sendFile(.)函数和代码 3-16 的 recvFile(.)函数,使得发送方和接收方可以使用大小不同的缓冲区。

第 4 章 Netty

Netty 是基于 Java NIO 的异步事件驱动的网络应用框架，官网是 https://netty.io/。Netty 提供了高层次的抽象来简化网络编程，使用 Netty 可以快速开发网络应用。Netty 的易用性、可用性已经得到多个商用项目的验证。Netty 的内部实现很复杂，但是上手使用却很快。本章对应的源代码为附件中的 netty 项目。

4.1 Netty 的使用模型

Netty 框架采用的是分层处理的思想，在 Netty 中每个处理层称为一个 ChannelHandler，多个相连的处理层就构成了 ChannelHandler 链，用于数据发送的 ChannelHandler 链称为出站链，用于数据接收的 ChannelHandler 链称为入站链。对于从网络底层接收到的字节序列，Netty 已将其封装成 ByteBuf，开发者把 ByteBuf 交给自定义的入站 ChannelHandler 链进行流水线式的处理，在流水线的末端得到业务数据，如字符串、对象、文件等，这种在入站时进行格式转换的 ChannelHandler 也称为解码器；当要发送业务数据时，如字符串、对象、文件等，开发者把业务数据交给自定义的出站 ChannelHandler 链进行流水线式的处理，在流水线的末端得到 ByteBuf 或者字节序列，然后 Netty 将完成网络底层的发送，这种在出站时进行格式转换的 ChannelHandler 也称为编码器。对于开发者来说，不需要关注网络底层的传输，只需要设计、搭建 ChannelHandler 链。当使用 Netty 时，开发者宜按照如图 4-1 所示的模式来设计入站链、出站链。

图 4-1 中带箭头的连线表示数据的流向，有些连线上注明了数据的类型，如 ByteBuf 或者业务数据。每个方块里的函数调用是驱动数据流动的方法。Netty 定义了接口 ChannelHandler 并提供了该接口的两个实现类，一个是表示入站处理的 ChannelInboundHandlerAdapter，另一个是表示出站处理的 ChannelOutboundHandlerAdapter。图 4-1 中所有的 InHander（包括 InHanderX 和 InHanderY）都是 ChannelInboundHandlerAdapter 的子类，所有的

OutHandler 都是 ChannelOutboundHandlerAdapter 的子类，见图 4-2，其中列出了在子类中需要重载的方法。

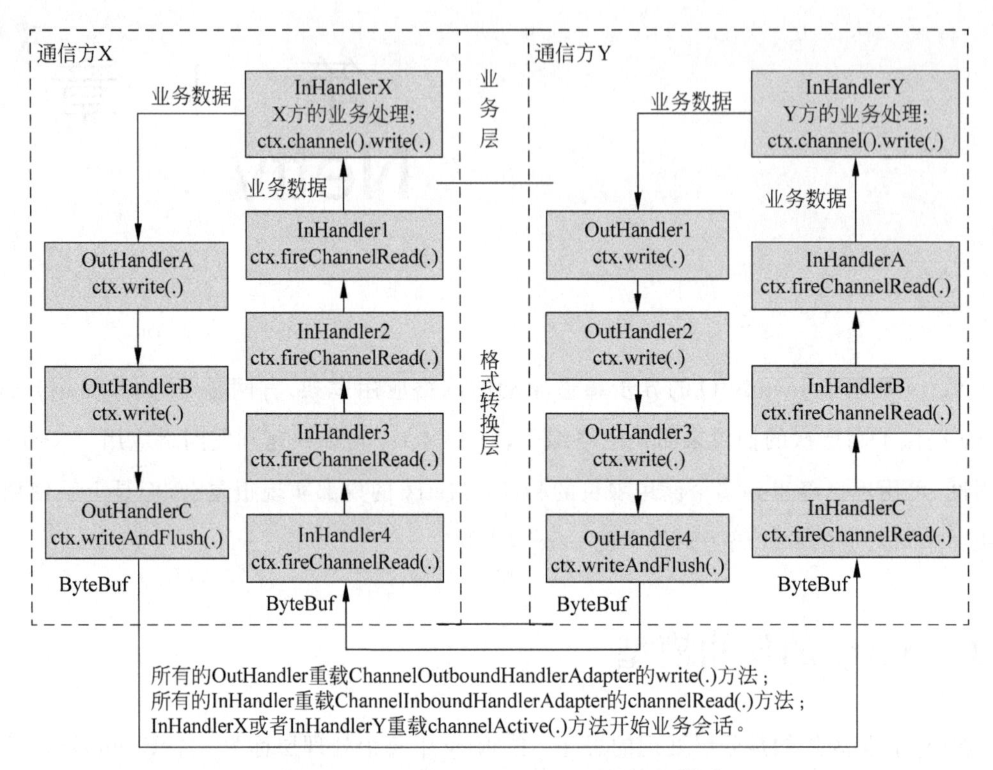

图 4-1　Netty 的 ChannelHandler 链使用模型

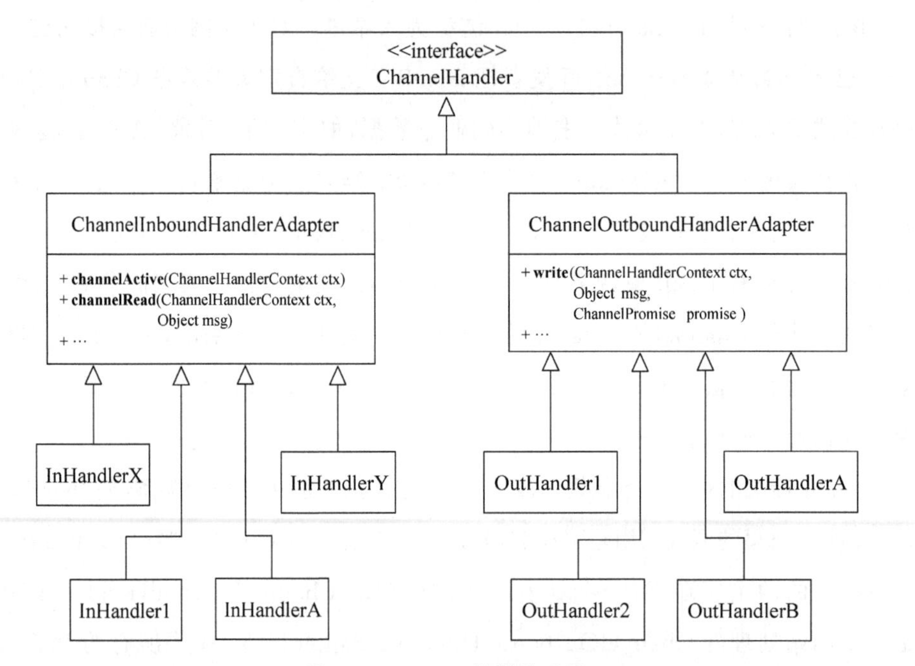

图 4-2　Handler 的类关系图

ChannelInboundHandlerAdapter 的子类重载父类的 channelRead(.)方法，在重载的 channelRead(.)方法中调用 ctx. fireChannelRead(.)将驱动入站数据流向入站链的下一个 InHandler，直至到达处理业务的 InHandlerX 或者 InHandlerY 的重载方法 channelRead(.)内。ChannelOutboundHandlerAdapter 的子类重载父类的 write(.)方法，在重载的 write(.)方法中调用 ctx. write(.)将驱动出站数据流向出站链的下一个 OutHandler，直至到达出站链最后一个 OutHandler 的重载方法 write(.)内，这时需要调用 ctx. writeAndFlush(.)才能将出站数据输出至网络底层并发送出去。在入站链末端是处理业务的 InHandlerX 或者 InHandlerY，在这两个 InHandler 的重载方法 channelRead(.)中，调用 ctx. channel(). write(.)或 ctx. pipeline(). write(.)将触发出站链顶端 OutHandler 的重载方法 write(.)的执行。当通信双方建立连接时，如果会话流程是以通信方 X 发送消息开始，则在 InHandlerX 的重载方法 channelActive(.)中调用 ctx. channel(). write(.)或 ctx. pipeline(). write(.)，该调用将触发通信方 X 出站链顶端 OutHandler 的重载方法 write(.)的执行；如果会话流程是以通信方 Y 发送消息开始，则在 InHandlerY 的重载方法 channelActive(.)中调用 ctx. channel(). write(.)或 ctx. pipeline(). write(.)，该调用将触发通信方 Y 出站链顶端 OutHandler 的重载方法 write(.)的执行。

4.2 Netty 的入站与出站

Netty 中入站信息的操作是靠事件驱动的。当下列某个入站事件发生时，ChannelInboundHandler 中对应的事件函数会自动执行。这些事件函数均是可重载的，ChannelInboundHandlerAdapter 对这些事件函数的默认实现就是调用对应的 ctx. fireXXX(.)把入站信息传递至下一个 ChannelInboundHandler。ChannelHandlerContext 和 ChannelPipeline 都有一系列的 fireXXX(.)方法，这两个类的 fireXXX(.)系列的同名方法在功能上并无区别，都是引发入站链中下一个 ChannelInboundHandler 的对应的事件函数的执行。

(1) 通道注册事件。自动执行事件函数 channelRegistered(.)，重载此事件函数时需手工调用 ctx. fireChannelRegistered(.)才能传递入站信息。

(2) 连接建立事件。自动执行事件函数 channelActive(.)，重载此事件函数时需手工调用 ctx. fireChannelActive(.)才能传递入站信息。

(3) 接收数据事件。自动执行事件函数 channelRead(.)，重载此事件函数时需手工调用 ctx. fireChannelRead(.)才能传递入站信息。

(4) 接收数据完成事件。自动执行事件函数 channelReadComplete(.)，重载此事件函数时需手工调用 ctx. fireChannelReadComplete(.)才能传递入站信息。

(5) 异常通知事件。自动执行事件函数 exceptionCaught(.),重载此事件函数时需手工调用 ctx. fireExceptionCaught(.)才能传递入站信息。

(6) 用户自定义事件。自动执行事件函数 userEventTriggered(.),重载此事件函数时需手工调用 ctx. fireUserEventTriggered(.)才能传递入站信息。

(7) 通道可写状态变化事件。自动执行事件函数 channelWritabilityChanged(.),重载此事件函数时需手工调用 ctx. fireChannelWritabilityChanged(.)才能传递入站信息。

(8) 连接关闭事件。自动执行事件函数 channelInactive(.),重载此事件函数时需手工调用 ctx. fireChannelInactive(.)才能传递入站信息。

Netty 中 ChannelOutboundHandler 提供下列出站操作,这些方法均是可重载的。

(1) 请求绑定端口 bind(.)。重载此操作函数时需手工调用 ctx. bind(.)才能传递出站信息。

(2) 请求连接远程节点 connect(.)。重载此操作函数时需手工调用 ctx. connect(.)才能传递出站信息。

(3) 请求发送数据 write(.)。重载此操作函数时需手工调用 ctx. write(.)才能传递出站信息。

(4) 请求冲刷数据 flush(.)。重载此操作函数时需手工调用 ctx. flush(.)才能传递出站信息。

(5) 请求发送并冲刷数据 writeAndFlush(.)。重载此操作函数时需手工调用 ctx. writeAndFlush(.)才能传递出站信息。

(6) 请求读取更多数据 read(.)。重载此操作函数时需手工调用 ctx. read(.)才能传递出站信息。

(7) 请求注销通道 deregister(.)。重载此操作函数时需手工调用 ctx. deregister(.)才能传递出站信息。

(8) 请求断开连接 disconnect(.)。重载此操作函数时需手工调用 ctx. disconnect(.)才能传递出站信息。

(9) 请求关闭通道 close(.)。重载此操作函数时需手工调用 ctx. close(.)才能传递出站信息。

上述这一组 9 个方法不但在 ChannelOutboundHandler 中有,在 ChannelHandlerContext 中也有同名的这组方法,在 Channel 和 ChannelPipeline 也有这组方法。使用方式通常是重载 ChannelOutboundHandler 的某个出站操作如 write(.),在重载函数中调用 ChannelHandlerContext 的同名方法如 ctx. write(.),因为 ctx. write(.)会引发出站链中下一个 ChannelOutboundHandler 的 write(.)方法的执行,这样就可以把出站操作一直传递下去。

此外，通常是在入站链末端的 ChannelInboundHandler 的重载函数 channelRead(.)、channelActive(.)中调用 ctx. channel(). write(.)或者 ctx. pipeline(). write(.)，这两个调用会触发出站链首端 ChannelOutboundHandler 的 write(.)的执行。其他方法同理。

Netty 的出站操作都是异步的(即不会等待操作的结果，而是在操作完成时的回调中得到结果)，因此这些异步操作通常需要设置回调的监听器。除了 read(.)和 flush(.)，这些方法都有一个 ChannelPromise 类型的参数 promise，该参数用来存放监听器(Listener)，当此出站操作真正完成(不论成功或失败)时，会自动执行监听器的 operationComplete(.)方法。设置监听器有两种途径。

途径 1：在 ChannelOutboundHandler 内重载出站操作时调用 promise. addListener(.)添加监听器。

途径 2：在入站链末端 ChannelInboundHandler 的重载函数 channelRead(.)、channelActive(.)内调用出站操作时，如调用 ctx. channel(). write(.)或者 ctx. pipeline(). write(.)时，可以采用 ChannelFuture 来注册监听器，代码如下。

```
ChannelFuture future = ctx.channel().write(msg);
future.addListener(new ChannelFutureListener() {
    @Override
    public void operationComplete(ChannelFuture future) throws Exception {
            ...
    }
});
```

这里的 future 将在出站链首端的 ChannelOutboundHandler 执行其 write(.)方法时自动作为 ChannelPromise 类型的参数传入。

不论采用上述何种方式设置监听器，在 ChannelOutboundHandler 重载的 write(.)方法内进行出站消息传递时，需要调用 ctx. write(msg, promise)将 promise 传入下一个 ChannelOutboundHandler 的 write(.)方法，而且需要在出站链中一直传递 ChannelPromise 参数，这样监听器才会生效。

4.3　服务端框架代码

使用 Netty 框架时服务端代码见代码 4-1，这是一个使用了“模板方法”设计模式的框架式服务端，具体和特定的会话功能只需在子类中重载其抽象方法即可。

代码 4-1：类 Server

```
1  public abstract class Server {
2      private int port;
```

```
3       protected Server(int port){
4           this.port = port;
5       }
6       void service(){
7           EventLoopGroup bossGroup = new NioEventLoopGroup();
8           EventLoopGroup workerGroup = new NioEventLoopGroup();
9           try {
10              ServerBootstrap b = new ServerBootstrap();
11              b.group(bossGroup, workerGroup)
12                      .channel(NioServerSocketChannel.class)
13                      .childHandler(new ChannelInitializer<SocketChannel>() {
14                          @Override
15                          public void initChannel(SocketChannel ch)
                            throws Exception {
16                              addHandlers(ch);
17                          }
18                      })
19                      .option(ChannelOption.SO_BACKLOG, 128)
20                      .childOption(ChannelOption.SO_KEEPALIVE, true);
21              ChannelFuture f = b.bind(this.port).sync();
22              System.out.println("服务器开启服务端口" + port + ",等待连接请求……");
23              //等待服务端口关闭
24              f.channel().closeFuture().sync();
25          }catch (Exception e){
26              e.printStackTrace();
27          } finally {
28              workerGroup.shutdownGracefully();
29              bossGroup.shutdownGracefully();
30          }
31      }
32      abstract void addHandlers(SocketChannel ch);
33  }
```

第7行和第8行的bossGroup和workerGroup是两个线程池，它们默认线程数为CPU核芯数乘以2，bossGroup用于接收客户端传过来的请求，接受请求后将后续操作交由workerGroup处理。第32行的抽象方法addHandlers(.)在第16行被调用，addHandlers(.)的具体功能将在Server的子类中实现，通常是把自定义的Handler实例添加到ChannelHandler链。

4.4 客户端框架代码

使用Netty框架时客户端代码见代码4-2，这是一个使用了“模板方法”设计模式的框架式客户端，具体和特定的会话功能只需在子类中重载其抽象方法即可。

代码 4-2：类 Client

```
1  public abstract class Client {
2      private String serverIP;
3      private int port;
4      protected Client(String ip, int port){
5          this.serverIP = ip;
6          this.port = port;
7      }
8      void communicate() {
9          EventLoopGroup workerGroup = new NioEventLoopGroup();
10         try {
11             Bootstrap b = new Bootstrap();
12             b.group(workerGroup);
13             b.channel(NioSocketChannel.class);
14             b.option(ChannelOption.SO_KEEPALIVE, true);
15             b.handler(new ChannelInitializer<SocketChannel>() {
16                 @Override
17                 public void initChannel(SocketChannel ch) throws Exception {
18                     addHandlers(ch);
19                 }
20             });
21             //客户端发起连接请求
22             ChannelFuture f = b.connect(this.serverIP, this.port).sync();
23             f.channel().closeFuture().sync();
24         }catch (Exception e){
25             e.printStackTrace();
26         } finally {
27             workerGroup.shutdownGracefully();
28         }
29     }
30     abstract void addHandlers(SocketChannel ch);
31 }
```

客户端启动过程基本和服务端的启动过程类似，在客户端只用到一个线程池。第 30 行的抽象方法 addHandlers(.)在第 18 行被调用，addHandlers(.)的具体功能将在 Client 的子类中实现，通常是把自定义的 Handler 实例添加到 ChannelHandler 链。

4.5 ByteBuf、分帧以及 ChannelHandler 链

Netty 定义了自己的缓冲区类型 ByteBuf。ByteBuf 内部维护了读位置和写位置两个值，这样就不必像 NIO 的 ByteBuffer 一样使用 flip()；在初始化 ByteBuf 时可以设置初始容量和最大容量，当往 Bytebuf 里写入数据超出初始容量时，ByteBuf 会自动扩容。图 4-3

显示了这几个值的关系。

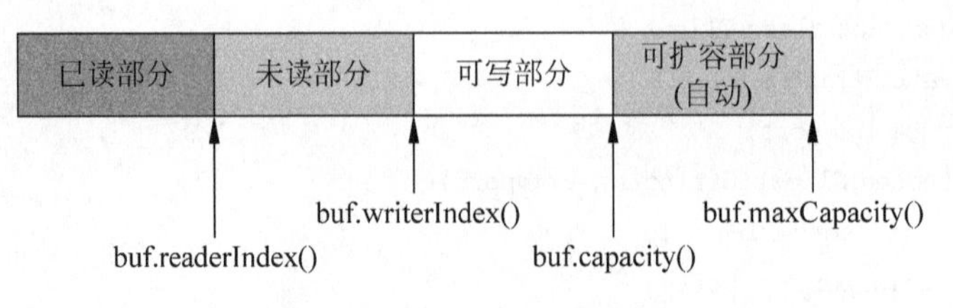

图 4-3 ByteBuf 的索引值

(1) ByteBuf 常用的创建函数如下。

Unpooled. buffer()——返回一个基于堆内存存储的 ByteBuf。

Unpooled. directBuffer()——返回一个基于直接内存存储的 ByteBuf。

Unpooled. wrappedBuffer(byte[])——返回一个包装了给定字节数组的 ByteBuf。

Unpooled. copiedBuffer(byte[])——返回一个复制了给定字节数组的 ByteBuf。

(2) ByteBuf 常用的读操作函数如下。

buf. readBytes(byte[] dst)——把 buf 里的数据全部读取到 dst 字节数组中,这个操作会自动修改 buf. readerIndex()值,这里 dst 数组的大小通常等于 buf. readableBytes()。

(3) ByteBuf 常用的写操作函数如下。

buf. writeBytes(byte[] src)——把字节数组 src 里面的数据全部写入 buf,这个操作会自动修改 buf. writerIndex()值。

使用 Netty 框架时,发送信息是主动的,而接收信息是被动的。通常情况下,开发者无须显式地调用通道的 read(.)方法来接收数据。当有信息传入时,会自动引发位于入站链首端(即最靠近网络底层)ChannelInboundHandler 的 channelRead(.)方法的执行,且执行此方法时第二个参数值正是封装了所传入字节序列的 ByteBuf 对象。因此,在接收数据方面,Netty 与之前 BIO、NIO 等的 read(.)操作有很大差别:使用 Netty 时,接收方并不能直接控制每次接收到的最多字节数,很可能一次接收到的信息超出了发送方的单帧内容,因此使用 Netty 也需要分帧。所幸的是,Netty 提供了处理基于长度的协议的解码器。基于长度的协议是通过将它的数据内容的长度编码到帧的头部来定义帧(也就是在第 1~3 章中常用的发送数据之前先发送数据的字节个数这种做法),Netty 提供的名为 LengthFieldBasedFrameDecoder 的解码器可以根据编码进帧头部中的长度值提取帧内容,长度值字段的偏移量和长度在构造器中指定。

在重载 Server 或者 Client 的 addHandlers(.)方法时,遵循如下两条规则可以搭建符合 Netty 标准的 Handler 流水线。

规则 1:调用函数. addLast(.),依次添加各 Handler 实例,优先添加距离网线近的。

规则 2:出站链与入站链不会相互干扰,可以添加完出站链的各项再添加入站链的各项，或者添加完入站链的各项再添加出站链的各项，甚至出站链与入站链的各项混杂着添加，无论何种添加顺序，不得违背规则 1。

假设某个通信者有如图 4-4 所示的出站链和入站链。

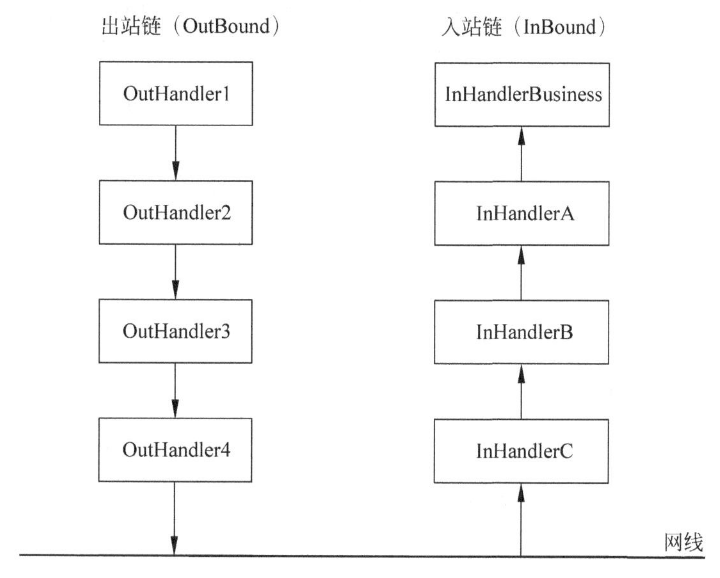

图 4-4　某通信者的出站链和入站链

针对图 4-4，构建 Handler 流水线时可供选择的添加顺序很多，下面只是其中之二。

添加顺序 1	添加顺序 2

void addHandlers(SocketChannel ch){
    ch.pipeline().addLast(new InHandlerC());
    ch.pipeline().addLast(new InHandlerB());
    ch.pipeline().addLast(new InHandlerA());
    ch.pipeline().addLast(
            new InHandlerBusiness());
    ch.pipeline().addLast(new OutHandler4());
    ch.pipeline().addLast(new OutHandler3());
    ch.pipeline().addLast(new OutHandler2());
    ch.pipeline().addLast(new OutHandler1());
}
	void addHandlers(SocketChannel ch){
	    ch.pipeline().addLast(new OutHandler4());
	    ch.pipeline().addLast(new InHandlerC());
	    ch.pipeline().addLast(new OutHandler3());
	    ch.pipeline().addLast(new OutHandler2());
	    ch.pipeline().addLast(new InHandlerB());
	    ch.pipeline().addLast(new OutHandler1());
	    ch.pipeline().addLast(new InHandlerA());
	    ch.pipeline().addlast(
	            new InHandlerBusiness());
	}

不论何种添加顺序，必须遵守上述两条规则。

4.6　案例 1：传输字符串的会话

使用 Netty 完成如图 4-5 所示的会话。

设计 ChannelHandler 链如图 4-6 所示。

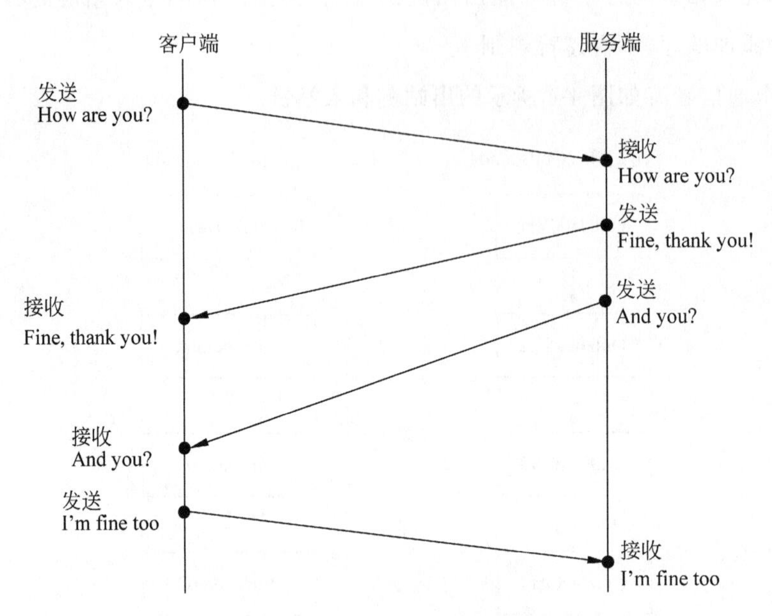

图 4-5　传输字符串的会话流程

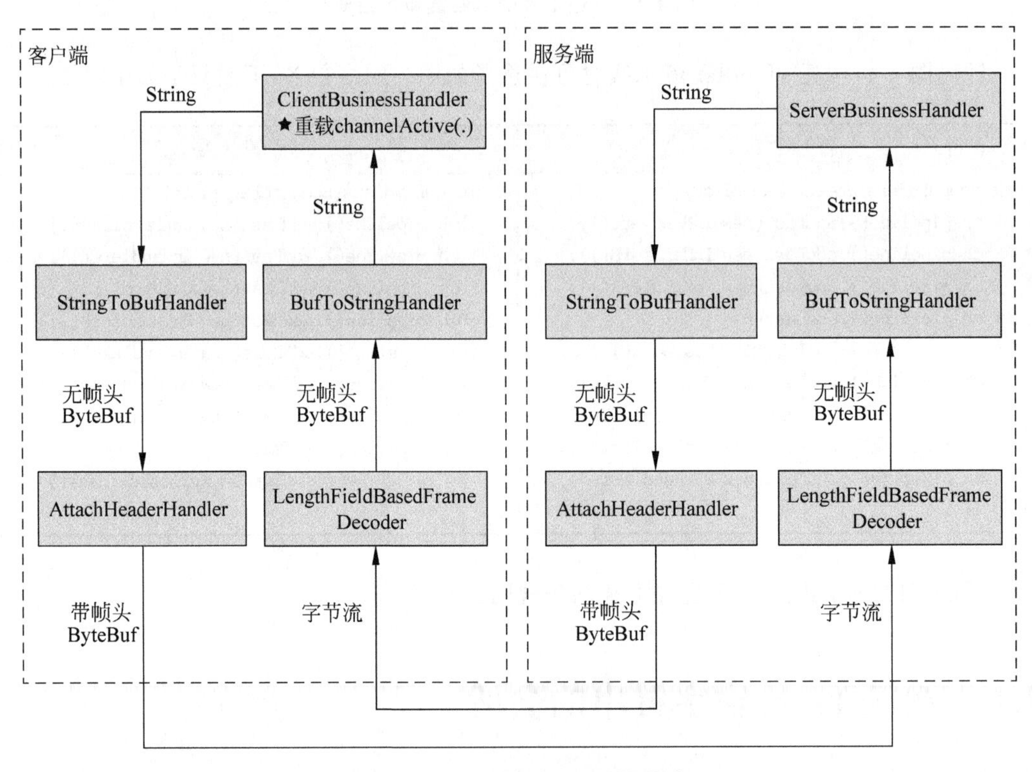

图 4-6　ChannelHandler 链的设计

首先导入 Netty 依赖,代码如下。

```
<dependency>
    <groupId>io.netty</groupId>
    <artifactId>netty-all</artifactId>
    <version>4.1.51.Final</version>
</dependency>
```

类 StringToBufHandler 参见代码 4-3。

代码 4-3:类 StringToBufHandler

```
1  @ChannelHandler.Sharable
2  public class StringToBufHandler extends ChannelOutboundHandlerAdapter {
3      @Override
4      public void write(ChannelHandlerContext ctx,
                         Object msg,
                         ChannelPromise promise) throws Exception {
5          String str = (String)msg;
6          promise.addListener(new ChannelFutureListener() {
7              @Override
8              public void operationComplete(ChannelFuture future) throws Exception {
9                  if (future.isSuccess()) {
10                     System.out.println("成功发送:" + str);
11                 }
12             }
13         });
14         byte[] bytes = str.getBytes(CharsetUtil.UTF_8);
15         ByteBuf buf = Unpooled.wrappedBuffer(bytes);
16         ReferenceCountUtil.release(msg);
17         ctx.write(buf, promise);
18     }
19 }
```

StringToBufHandler 是出站链的首端,因此在其 write(.)方法中第 6 行向参数 promise 注册监听器,用于在发送操作真正完成时给出第 10 行的提示信息。为了让第 6 行添加的监听器最终生效,必须在第 17 行向下一个 ChannelOutboundHandler 传递出站数据时把 promise 也传入,并且出站链中后续的每个 ctx.write(.)都必须传入 promise。

因为参数 msg 不被传递至下一个 ChannelOutboundHandler,所以这里务必要手动调用 ReferenceCountUtil.release(msg)来释放该参数以避免内存泄漏。其他 ChannelHandler 也要注意手动释放不被传递的参数。

类 AttachHeaderHandler 参见代码 4-4。

代码 4-4：类 AttachHeaderHandler

```
1  @ChannelHandler.Sharable
2  public class AttachHeaderHandler extends ChannelOutboundHandlerAdapter {
3      @Override
4      public void write(ChannelHandlerContext ctx,
                         Object msg,
                         ChannelPromise promise) throws Exception {
5          ByteBuf buf = (ByteBuf)msg;
6          ByteBuf countBuf = Unpooled.buffer(4,4);
7          countBuf.writeInt(buf.capacity());
8          ByteBuf newBuf = Unpooled.wrappedBuffer(countBuf,buf);
9          ctx.writeAndFlush(newBuf, promise);
10     }
11 }
```

与代码 4-3 的参数 msg 不同，代码 4-4 的参数 msg 被进一步封装到第 8 行的 ByteBuf 里并在第 9 行被传递，所以这里不能手动调用.release(.)来释放参数 msg。

AttachHeaderHandler 位于出站链的末端，因此第 9 行必须调用 ctx.writeAndFlush(.)而不能调用 ctx.write(.)，这样才能把出站数据冲刷至网络底层。另外，第 9 行传入了第 2 个参数 promise，这是为了让在出站链首端添加的监听器生效。

类 BufToStringHandler 参见代码 4-5。

代码 4-5：类 BufToStringHandler

```
1  @ChannelHandler.Sharable
2  public class BufToStringHandler extends ChannelInboundHandlerAdapter {
3      @Override
4      public void channelRead(ChannelHandlerContext ctx,
                               Object msg) throws Exception {
5          ByteBuf buf = (ByteBuf)msg;
6          String str = buf.toString(CharsetUtil.UTF_8);
7          ctx.fireChannelRead(str);
8          ReferenceCountUtil.release(msg);
9      }
10 }
```

这里因为参数 msg 的内容在第 6 行被转换至字符串，而且在第 7 行进一步传递入站数据时没有用到 msg，所以需要在第 8 行手动释放 msg 对象。

类 ServerBusinessHandler 参见代码 4-6。

代码 4-6：类 ServerBusinessHandler

```
1  @ChannelHandler.Sharable
2  public class ServerBusinessHandler extends ChannelInboundHandlerAdapter {
```

```
3       @Override
4       public void channelActive(ChannelHandlerContext ctx) throws Exception {
5           System.out.println("服务端建立连接,开始会话……");
6       }
7       int stepIndex = 0;
8       @Override
9       public void channelRead(ChannelHandlerContext ctx,
                                    Object msg) throws Exception {
10          String str = (String)msg;
11          System.out.println("成功接收: " + str);
12          ReferenceCountUtil.release(msg);
13          stepIndex++;
14          if (stepIndex == 1){
15              str = "I am fine, thank you!";
16              ctx.channel().write(str);
17              str = "And you?";
18              ctx.channel().write(str);
19          }
20          if (stepIndex == 2){
21              ctx.channel().close();
22              System.out.println("会话结束,服务端关闭连接!");
23          }
24      }
25 }
```

这是一个入站 Handler，当第 9 行的 channelRead(.)函数被接收数据事件触发执行时，数据接收已经完成，而参数 msg 正是所接收到的数据，所以直接在第 11 行显示提示信息。第 4 行重载 channelActive(.)只是在双方建立连接时输出提示信息。根据会话流程，服务端第一次接收信息后是两次连续的发送，见第 16 行和第 18 行，第二次接收信息后关闭通道结束会话，见第 21 行。

类 ClientBusinessHandler 参见代码 4-7。

代码 4-7：类 ClientBusinessHandler

```
1  @ChannelHandler.Sharable
2  public class ClientBusinessHandler extends ChannelInboundHandlerAdapter {
3      int stepIndex = 0;
4      @Override
5      public void channelActive(ChannelHandlerContext ctx) throws Exception {
6          System.out.println("客户端建立连接,开始会话……");
7          String s = "How are you?";
8          ctx.channel().write(s);
9      }
10     @Override
```

```
11      public void channelRead(ChannelHandlerContext ctx,
                                      Object msg) throws Exception {
12          String str = (String)msg;
13          System.out.println("成功接收：" + str);
14          ReferenceCountUtil.release(msg);
15          stepIndex++;
16          if (stepIndex == 2){
17              str = "I'm fine, too!";
18              ChannelFuture future = ctx.channel().write(str);
19              future.addListener(new ChannelFutureListener() {
20                  @Override
21                  public void operationComplete(ChannelFuture future)
                    throws Exception {
22                      if (future.isSuccess()){
23                          ctx.channel().close();
24                          System.out.println("会话结束,客户端关闭连接!");
25                      }
26                  }
27              });
28          }
29      }
30 }
```

因为双方建立连接后客户端首先发送信息，故第 5 行重载 channelActive(.)以便客户端发送会话的第 1 条信息。

第 23 行是客户端主动关闭通道，对比代码 4-6 的第 21 行服务端主动关闭通道。一般来说，主动关闭通道发生在读、写操作完毕的时候。读操作是入站事件，故发生在 ChannelInboundHandler 里，写操作是出站事件，可能会在 ChannelInboundHandler 里调用 ctx.channel().write(.)，但最终还是发生在 ChannelOutboundHandler 里。在 ChannelInboundHandler 里很容易判断读操作是否完毕，因为 Netty 里的“读”是自动完成的，当事件处理过程 channelRead()被触发执行时表示读操作已经结束。代码 4-6 的第 21 行只需直接关闭通道，因为此时会话流程中服务器的最后一次通信的“读”操作必然已经结束；但是 Netty 里的“写”操作需要主动调用并且是异步的，因此只能使用监听器，在写操作完成时将自动执行监听器里的操作。因为会话流程中客户端的最后一次通信是“写”，程序必须确保写操作已经完成才能关闭通道，所以在第 19 行注册监听器，在第 21 行的 operationComplete(.)回调函数里稳妥地关闭通道并提示会话结束。调用 ctx.close()或者调用 ctx.channel().close()都可以关闭通道，两种方法效果相同。另外，注意代码 4-7 的第 19 行与代码 4-3 的第 6 行两种添加监听器的方式及效果，代码 4-7 的第 19 行这种方式采用 ChannelFuture

添加的监听器只对第18行的这一次写操作有效，而代码4-3的第6行这种方式于出站链首端添加的监听器对每一次的 ctx. channel(). write(.)或者 ctx. pipeline(). write(.)写操作都有效。

服务端启动类参见代码4-8，继承了 Server 类。

代码 4-8：类 Server1

```
1  public class Server1 extends Server{
2      public static void main(String[ ] args) {
3          Server1 server = new Server1();
4          server.service();
5      }
6      public Server1(){
7          super(8001);
8      }
9      @Override
10     void addHandlers(SocketChannel ch) {
11         ch.pipeline().addLast(
                   new LengthFieldBasedFrameDecoder(Integer.MAX_VALUE, 0, 4, 0, 4));
12         ch.pipeline().addLast(new BufToStringHandler());
13         ch.pipeline().addLast(new ServerBusinessHandler());
14         ch.pipeline().addLast(new AttachHeaderHandler());
15         ch.pipeline().addLast(new StringToBufHandler());
16     }
17 }
```

参照图4-6所示的服务端，在重载的 addHandlers(.)里添加 ChannelHandler。

客户端启动类参见代码4-9，继承了 Client 类。

代码 4-9：类 Client1

```
1  public class Client1 extends Client{
2      public static void main(String[ ] args) {
3          ExecutorService fixPool = Executors.newCachedThreadPool();
4          for (int i = 0; i < 3; i++) {
5              fixPool.execute(
6                      () -> {
7                          try {
8                              Client1 client = new Client1();
9                              client.communicate();
10                         } catch (Exception ex) {
11                             ex.printStackTrace();
12                         }
```

```
13            }
14        );
15        }
16        fixPool.shutdown();
17    }
18    public Client1(){
19        super("127.0.0.1", 8001);
20    }
21    @Override
22    void addHandlers(SocketChannel ch) {
23        ch.pipeline().addLast(
              new LengthFieldBasedFrameDecoder(Integer.MAX_VALUE, 0, 4, 0, 4));
24        ch.pipeline().addLast(new BufToStringHandler());
25        ch.pipeline().addLast(new ClientBusinessHandler());
26        ch.pipeline().addLast(new AttachHeaderHandler());
27        ch.pipeline().addLast(new StringToBufHandler());
28    }
29 }
```

参照图 4-6 所示的客户端，在重载的 addHandlers(.)中添加 ChannelHandler。用多线程模拟多客户端的并发访问。运行时，先启动附件代码 netty 项目的 Server1，再启动附件代码 netty 项目的 Client1。服务端及客户端控制台分别显示如下信息。

服务端控制台信息	客户端控制台信息
服务器开启服务端口 8001，等待连接请求…… 服务端建立连接，开始会话…… 成功接收：How are you? 成功发送：I am fine，thank you! 成功发送：And you? 成功接收：I'm fine，too! 会话结束，服务端关闭连接！	客户端建立连接，开始会话…… 成功发送：How are you? 成功接收：I am fine，thank you! 成功接收：And you? 成功发送：I'm fine，too! 会话结束，客户端关闭连接！

4.7 案例 2：传输对象的会话

当客户端与服务端连接成功时，它们之间进行如图 4-7 所示的会话。

设计 ChannelHandler 链如图 4-8 所示。

类 AttachHeaderHandler 在前一节用过，见代码 4-4。

类 Serializer 见代码 4-10。

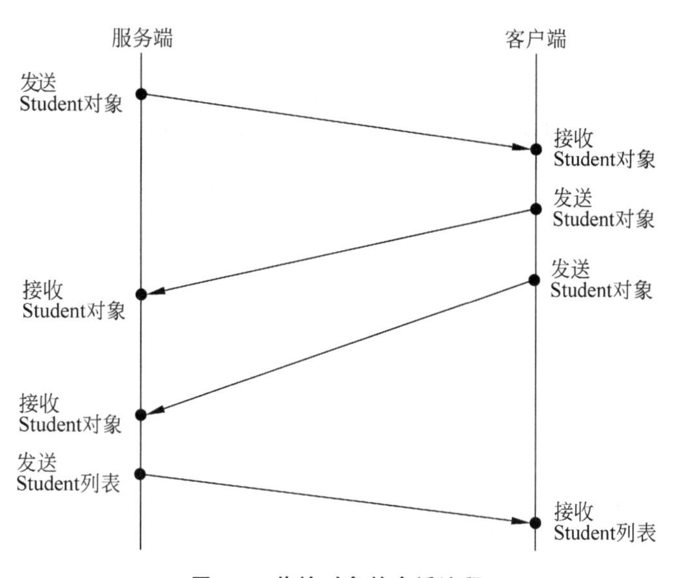

图 4-7 传输对象的会话流程

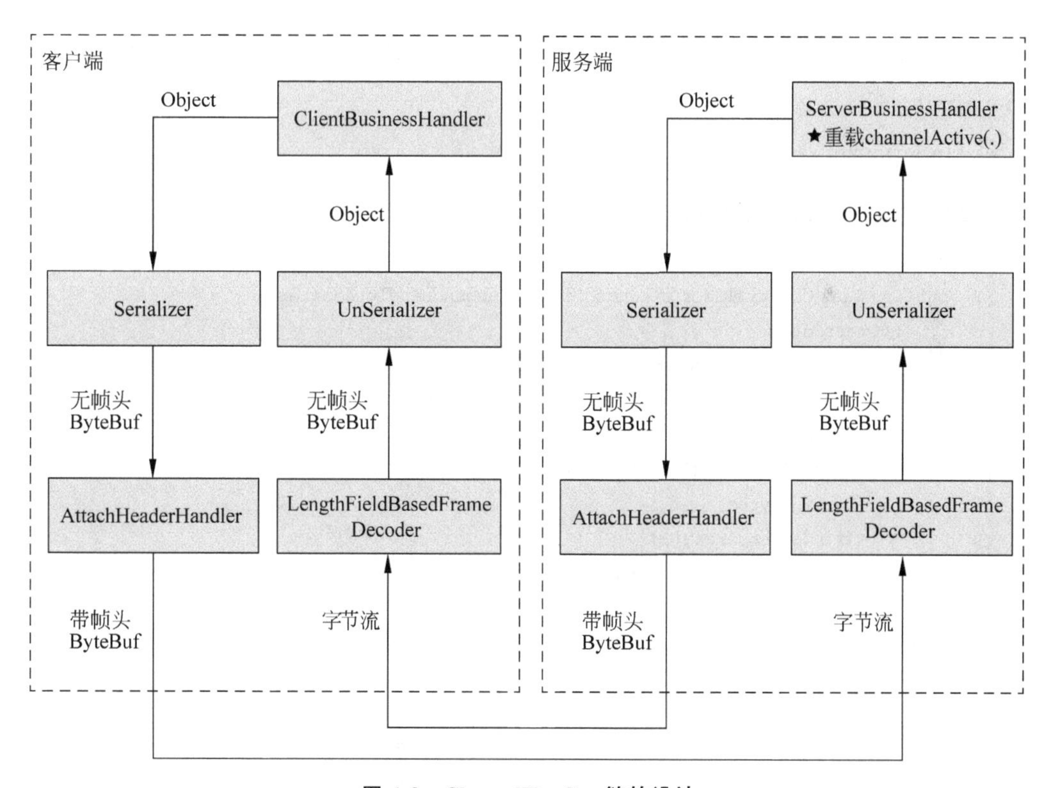

图 4-8 ChannelHandler 链的设计

代码 4-10：类 Serializer

```
1  @ChannelHandler.Sharable
2  public class Serializer extends ChannelOutboundHandlerAdapter {
3      @Override
```

```
4       public void write(ChannelHandlerContext ctx,
                          Object msg,
                          ChannelPromise promise) throws Exception {
5           promise.addListener(new ChannelFutureListener() {
6               @Override
7               public void operationComplete(ChannelFuture future) throws Exception {
8                   if (future.isSuccess()) {
9                       System.out.println("成功发送：" + msg);
10                      ReferenceCountUtil.release(msg);
11                  }
12              }
13          });
14          //把对象转成字节数组
15          byte[] bytes = Utils.objSerializableToByteArray(msg);
16          ByteBuf buf = Unpooled.wrappedBuffer(bytes);
17          ctx.write(buf, promise);
18      }
19 }
```

第15行的函数objSerializableToByteArray(.)见代码1-14。这里要求所传输的对象具有Serializable接口。

类UnSerializer见代码4-11。

代码4-11：类UnSerializer

```
1  @ChannelHandler.Sharable
2  public class UnSerializer extends ChannelInboundHandlerAdapter {
3      @Override
4      public void channelRead(ChannelHandlerContext ctx,
                               Object msg) throws Exception {
5          //从ByteBuf->byte数组->对象
6          ByteBuf buf = (ByteBuf)msg;
7          byte[] bytes;
8          if (buf.hasArray()){
9              bytes = buf.array();
10         }else{
11             bytes = new byte[buf.readableBytes()];
12             buf.readBytes(bytes);
13         }
14         Object obj = Utils.byteArrayToObjSerializable(bytes);
15         ctx.fireChannelRead(obj);
16         ReferenceCountUtil.release(msg);
17     }
18 }
```

第14行的函数byteArrayToObjSerializable(.)见代码1-15。

类ServerBusinessHandler参见代码4-12。

代码 4-12：类 ServerBusinessHandler

```
1  @ChannelHandler.Sharable
2  public class ServerBusinessHandler extends ChannelInboundHandlerAdapter {
3      int stepIndex = 0;
4      @Override
5      public void channelActive(ChannelHandlerContext ctx) throws Exception {
6          System.out.println("服务端建立连接,开始会话……");
7          Student stu = StuRepository.getStudent();
8          ctx.channel().write(stu);
9      }
10     @Override
11     public void channelRead(ChannelHandlerContext ctx,
                               Object msg) throws Exception {
12         Student st = (Student) msg;
13         System.out.println("成功接收: " + st);
14         ReferenceCountUtil.release(msg);
15         stepIndex++;
16         if (stepIndex == 2){
17             List<Student> list = StuRepository.getStudents();
18             ChannelFuture future = ctx.channel().write(list);
19             future.addListener(new ChannelFutureListener() {
20                 @Override
21                 public void operationComplete(ChannelFuture future)
                   throws Exception {
22                     if (future.isSuccess()){
23                         ctx.channel().close();
24                         System.out.println("会话结束,服务端关闭连接!");
25                     }
26                 }
27             });
28         }
29     }
30 }
```

由于写操作是异步的,所以第 23 行关闭通道的操作放到监听器里执行。

类 ClientBusinessHandler 参见代码 4-13。

代码 4-13：类 ClientBusinessHandler

```
1  public class ClientBusinessHandler extends ChannelInboundHandlerAdapter {
2      int stepIndex = 0;
3      @Override
4      public void channelActive(ChannelHandlerContext ctx) throws Exception {
5          System.out.println("客户端建立连接,开始会话……");
6      }
7      @Override
```

```
8       public void channelRead(ChannelHandlerContext ctx,
                                Object msg) throws Exception {
9           stepIndex++;
10          //第一次接收
11          if (stepIndex == 1) {
12              Student st = (Student) msg;
13              System.out.println("成功接收：" + st);
14              ReferenceCountUtil.release(msg);
15              st = StuRepository.getStudent();
16              ctx.channel().write(st);
17              st = StuRepository.getStudent();
18              ctx.channel().write(st);
19          }
20          if (stepIndex == 2){
21              List<Student> stList = (List<Student>) msg;
22              System.out.println("成功接收：" + stList);
23              ReferenceCountUtil.release(msg);
24              ctx.channel().close();
25              System.out.println("会话结束,客户端关闭连接!");
26          }
27      }
28 }
```

服务端启动类参见代码 4-14，继承了 Server 类。

代码 4-14：类 Server2

```
1  public class Server2 extends Server{
2      public static void main(String[] args) {
3          Server2 server = new Server2();
4          server.service();
5      }
6      public Server2(){
7          super(8001);
8      }
9      @Override
10     void addHandlers(SocketChannel ch) {
11         ch.pipeline().addLast(
                   new LengthFieldBasedFrameDecoder(Integer.MAX_VALUE, 0, 4, 0, 4));
12         ch.pipeline().addLast(new UnSerializer());
13         ch.pipeline().addLast(new ServerBusinessHandler());
14         ch.pipeline().addLast(new AttachHeaderHandler());
15         ch.pipeline().addLast(new Serializer());
16     }
17 }
```

参照如图 4-8 所示的服务端，在重载的 addHandlers(.)中添加 ChannelHandler。

客户端启动类参见代码 4-15，继承了 Client 类。

代码 4-15：类 Client2

```
1  public class Client2 extends Client{
2      public static void main(String[] args) {
3          ExecutorService fixPool = Executors.newCachedThreadPool();
4          for (int i = 0; i < 1; i++) {
5              fixPool.execute(
6                      () -> {
7                          try {
8                              Client2 client = new Client2();
9                              client.communicate();
10                         } catch (Exception ex) {
11                             ex.printStackTrace();
12                         }
13                     }
14             );
15         }
16         fixPool.shutdown();
17     }
18     public Client2(){
19         super("127.0.0.1", 8001);
20     }
21     @Override
22     void addHandlers(SocketChannel ch) {
23         ch.pipeline().addLast(
                   new LengthFieldBasedFrameDecoder(Integer.MAX_VALUE, 0, 4, 0, 4));
24         ch.pipeline().addLast(new UnSerializer());
25         ch.pipeline().addLast(new ClientBusinessHandler());
26         ch.pipeline().addLast(new AttachHeaderHandler());
27         ch.pipeline().addLast(new Serializer());
28     }
29 }
```

参照如图 4-8 所示的客户端，在重载的 addHandlers(.)中添加 ChannelHandler。用多线程模拟多客户端的并发访问。

运行时，先启动附件代码 netty 项目的 Server2，再启动附件代码 netty 项目的 Client2。

4.8 案例 3：传输文件的会话

当客户端与服务端连接成功时，它们之间进行如图 4-9 所示的会话。

设计 ChannelHandler 链如图 4-10 所示。

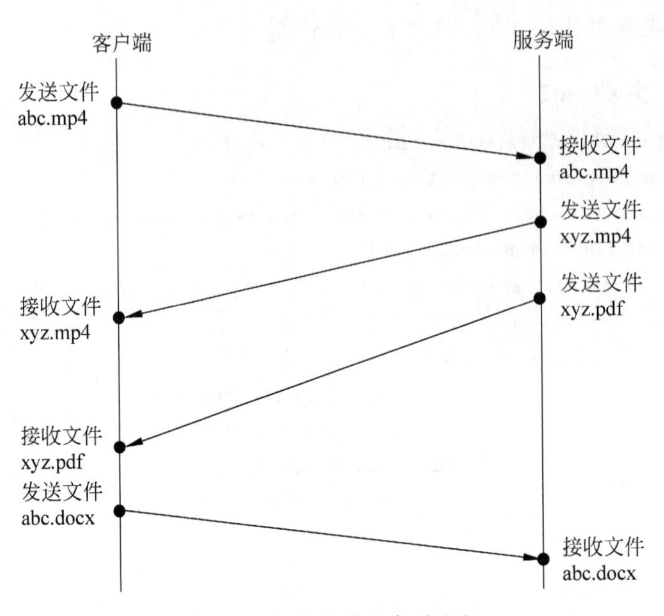

图 4-9 传输文件的会话流程

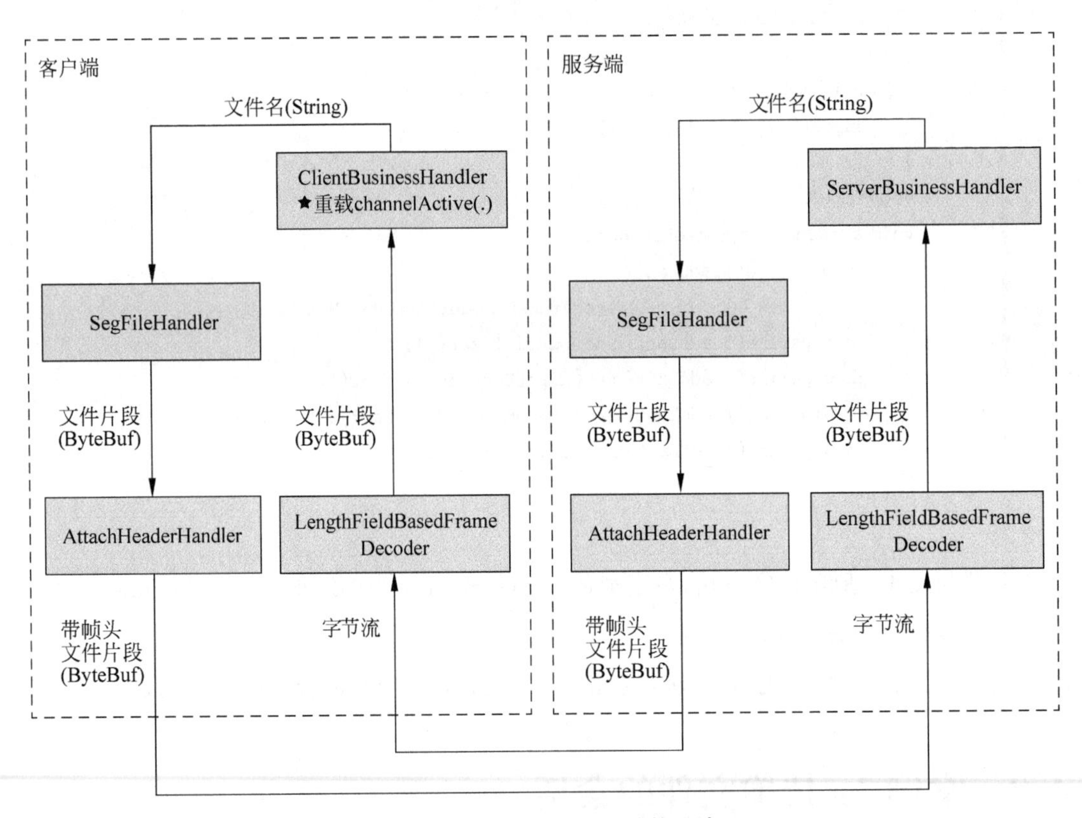

图 4-10 ChannelHandler 链的设计

对于大文件的传输，依然采用分片传输的方式，这与之前传输大文件的思路是一样的。在 Netty 的 Handler 链框架下，传输大文件的处理流程如图 4-11 所示。

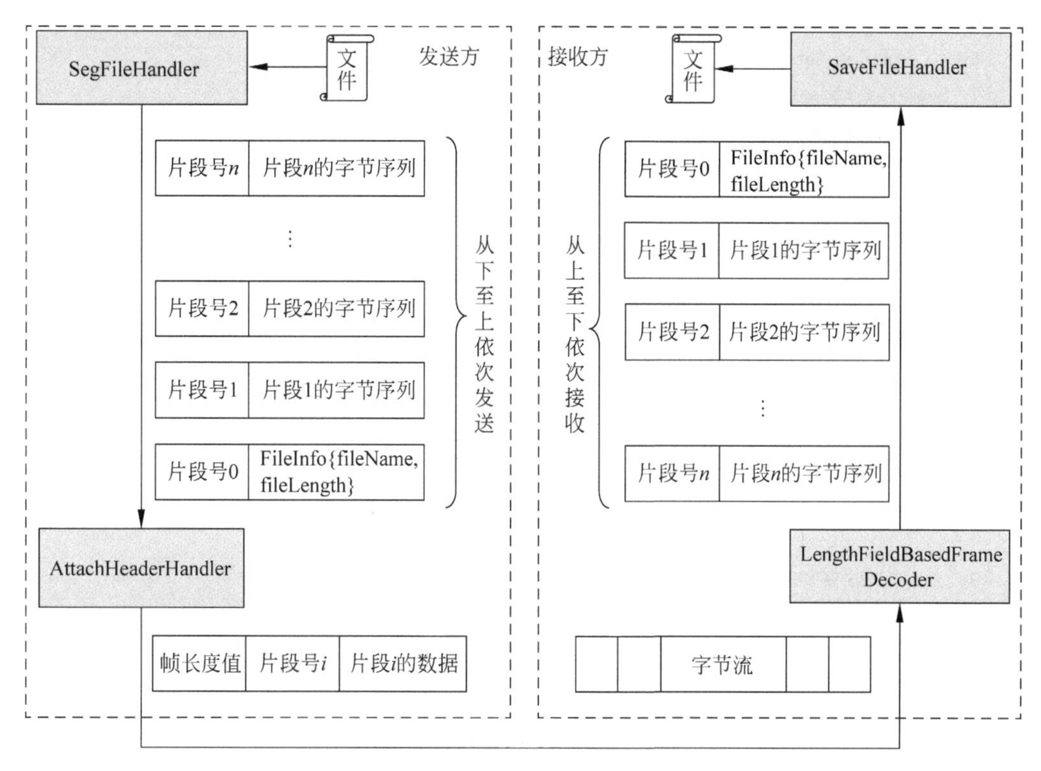

图 4-11 传输大文件的处理流程

类 SegFileHandler 从文件中读取数据片段，在其头部增加片段号构成数据帧依次传递给 AttachHeaderHandler，AttachHeaderHandler 对接收到的每一帧在其头部增加该帧长度值然后发送，接收方使用 LengthFieldBasedFrameDecoder 依据帧长度值对传入的字节流进行分帧，将拆解出来的每个数据帧（由片段号和数据构成）传递给 SaveFileHandler，SaverFileHandler 剔除片段号后再将每块数据依次写入文件。

定义类 FileInfo（代码 4-16）及 SendCmd（代码 4-17）作为类 SegFileHandler 的辅助。

代码 4-16：类 FileInfo

```
1  @Data
2  @AllArgsConstructor
3  public class FileInfo implements Serializable {
4      String fileName;
5      long fileLength;
6  }
```

代码 4-17：类 SendCmd

```
1  @Data
2  @AllArgsConstructor
3  public class SendCmd {
4      private String filePath;
```

```
5       private boolean closeAfterSending;
6       public SendCmd(String filePath){
7           this.filePath = filePath;
8           closeAfterSending = false;
9       }
10 }
```

类 SegFileHandler 见代码 4-18。

代码 4-18：类 SegFileHandler

```
1  public class SegFileHandler extends ChannelOutboundHandlerAdapter {
2       @Override
3       public void write(ChannelHandlerContext ctx,
                            Object msg,
                            ChannelPromise promise) throws Exception {
4           final SendCmd sendCmd = (SendCmd)msg;
5           String fPath = sendCmd.getFilePath();
6           System.out.println("开始发送文件：" + fPath);
7           String fName = fPath.substring(fPath.lastIndexOf("\\") + 1);
8           File f = new File(fPath);
9           long countBytesToSend = f.length();
10          FileInputStream fin = new FileInputStream(fPath);
11          FileChannel fcin = fin.getChannel();
12          FileInfo fInfo = new FileInfo(fName,countBytesToSend);
13          int bufSize = 1024 * 1024;
14          if (bufSize > countBytesToSend){
15              bufSize = (int)countBytesToSend;
16          }
17          ByteBuffer buffer = ByteBuffer.allocate(bufSize);
18          int segmentNo = 0;
19          //将 segmentNo 和 fInfo 组成 ByteBuf,作为第 0 块发送
20          byte[] bytes = Utils.objSerializableToByteArray(fInfo);
21          ByteBuf buf = Unpooled.buffer(4 + bytes.length,4 + bytes.length);
22          buf.writeInt(segmentNo);
23          buf.writeBytes(bytes);
24          ctx.write(buf);
25          segmentNo++;
26          //计算片段的总块数
27          ToLongBiFunction<Long, Long> function =
                                (a, b) -> {return a % b == 0 ? (a/b):(a/b) + 1;};
28          final long segmentCount =
                                function.applyAsLong(countBytesToSend, (long)bufSize);
29          //每块文件片段发送完成后将执行此公用监听器
30          ChannelFutureListener listener = new ChannelFutureListener() {
31              long restSegments = segmentCount;
32              @Override
33              public void operationComplete(ChannelFuture future)
                throws Exception {
```

```
34                  if (future.isSuccess()) {
35                      System.out.println(fPath + " 第" +
                        (segmentCount - restSegments + 1) + "块文件片段 发送成功!");
36                      restSegments--;
37                      if (restSegments == 0) {
38                          System.out.println("文件发送完成!" + fPath);
39                          //如果需要关闭通道
40                          if (sendCmd.isCloseAfterSending()) {
41                              ctx.channel().close();
42                              System.out.println("会话结束,客户端关闭连接!");
43                          }
44                      }
45                  }
46              }
47          };
48          //循环发送文件的多个片段
49          while (countBytesToSend > 0){
50              if (bufSize > countBytesToSend){
51                  bufSize = (int)countBytesToSend;
52                  buffer = ByteBuffer.allocate(bufSize);
53              }
54              //写模式的 buffer
55              buffer.clear();
56              //从文件读入数据至缓冲区,读满缓冲区
57              while (buffer.hasRemaining()){
58                  fcin.read(buffer);
59              }
60              buffer.flip();                          //读模式的 buffer
61              //将 segmentNo 和 文件缓冲区组成 ByteBuf,发送
62              buf = Unpooled.buffer(4 + buffer.capacity(),4 + buffer.capacity());
63              buf.writeInt(segmentNo);
64              buf.writeBytes(buffer);
65              //发送一块文件片段,当发送完成时回调 listener
66              ChannelFuture future = ctx.write(buf);
67              future.addListener(listener);
68              //因为 ctx.write(.)是异步操作,不会阻塞等待
69              //随即开始下一块文件片段的发送
70              countBytesToSend -= buffer.capacity();
71              segmentNo++;
72          }
73          //关闭文件
74          fcin.close();
75          fin.close();
76      }
77  }
```

第24行可以为write(.)操作添加监听器,改成如下代码。

```
ctx.write(buf).addListener(new ChannelFutureListener() {
    @Override
    public void operationComplete(ChannelFuture future) throws Exception {
        if (future.isSuccess()){
            System.out.println(fPath + " 第0号帧 发送成功!");
        }
    }
});
```

第27行使用了函数式接口ToLongBiFunction来计算片段的总块数。

因为ctx.write(.)是异步的,因此第62行至第64行不使用如下代码来构造buf,以避免可能发生的ByteBuf访问冲突。

```
//不用wrap方式构造buf
ByteBuffer segNoBuf = ByteBuffer.allocate(4);
segNoBuf.putInt(0,segmentNo);
buf = Unpooled.wrappedBuffer(segNoBuf, buffer);
```

第66行和第67行是为文件每块数据片段的发送指定回调的监听器,第33行的回调函数将检查每块数据片段是否发送成功。

类AttachHeaderHandler在4.6节用过,见代码4-4。

类SaveFileHandler的继承关系参见图4-12。

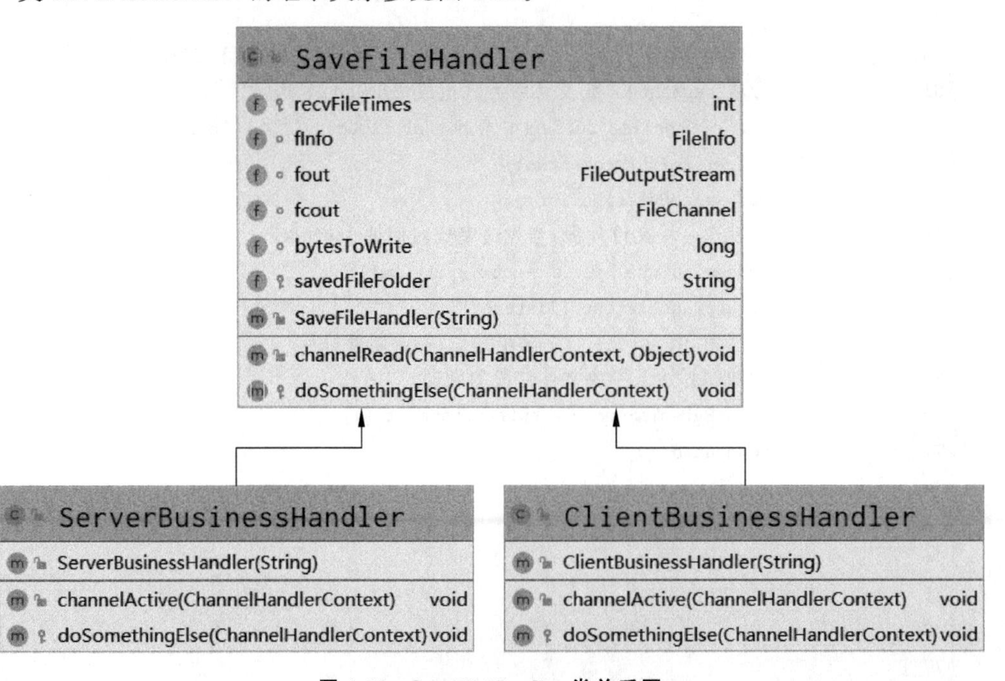

图4-12 SaveFileHandler类关系图

类 SaveFileHandler 参见代码 4-19。

代码 4-19：类 SaveFileHandler

```
1   abstract public class SaveFileHandler extends ChannelInboundHandlerAdapter {
2       protected int recvFileTimes = 0;      //统计接收文件的次数
3       FileInfo fInfo;
4       FileOutputStream fout;
5       FileChannel fcout;
6       long bytesToWrite;
7       protected String savedFileFolder;
8       public SaveFileHandler(String savedFileFolder){
9           this.savedFileFolder = savedFileFolder;
10      }
11      @Override
12      public void channelRead(ChannelHandlerContext ctx, Object msg)
        throws Exception {
13          ByteBuf buf = (ByteBuf)msg;
14          int segmentNo = buf.readInt();
15          byte[] bytes = new byte[buf.readableBytes()];
16          buf.readBytes(bytes);
17          ByteBuffer buffer = ByteBuffer.wrap(bytes);
18          ReferenceCountUtil.release(msg);
19          if (segmentNo == 0){
20              fInfo = (FileInfo) Utils.byteArrayToObjSerializable(bytes);
21              String fName = fInfo.getFileName();
22              bytesToWrite = fInfo.getFileLength();
23              fInfo.setFileName(savedFileFolder +
                            new Random().nextInt(10) + "-" + fName);
24              System.out.println("开始接收文件：" + fName);
25              fout = new FileOutputStream(new File(fInfo.getFileName()));
26              fcout = fout.getChannel();
27              return;
28          }
29          //数据块的处理
30          while (buffer.hasRemaining()){
31              fcout.write(buffer);
32          }
33          bytesToWrite -= buffer.capacity();
34          if (bytesToWrite == 0){
35              fcout.close();
36              fout.close();
37              System.out.println("保存文件完成!" + fInfo.getFileName());
38              recvFileTimes++;
39              doSomethingElse(ctx);
40          }
```

```
41      }
42      abstract protected void doSomethingElse(ChannelHandlerContext ctx);
43 }
```

类 SaveFileHandler 的重载方法 channelRead(.)接收包含段号的数据段，取出数据块并将其转存至文件中；另外采用了“模板方法”设计模式，在第 39 行被调用的抽象方法 doSomethingElse(.)在子类中的重载将实现特定的会话流程。

类 ServerBusinessHandler 参见代码 4-20。

代码 4-20：类 ServerBusinessHandler

```
1  public class ServerBusinessHandler extends SaveFileHandler {
2      public ServerBusinessHandler(String savedFilePath) {
3          super(savedFilePath);
4      }
5      @Override
6      public void channelActive(ChannelHandlerContext ctx) throws Exception {
7          System.out.println("服务端建立连接，开始会话……");
8      }
9      @Override
10     protected void doSomethingElse(ChannelHandlerContext ctx) {
11         //这里只需进行文件的发送及关闭通道
12         //文件的接收在父类的 channelRead(.)中完成
13         if(this.recvFileTimes == 1) {
14             String fPath = this.savedFileFolder + "xyz.mp4";
15             ctx.channel().write(new SendCmd(fPath));
16             fPath = this.savedFileFolder + "xyz.pdf";
17             ctx.channel().write(new SendCmd(fPath));
18         }
19         if (this.recvFileTimes == 2) {
20             //在读操作之后的通道关闭可直接进行
21             ctx.channel().close();
22             System.out.println("会话结束，服务端关闭连接!");
23         }
24     }
25 }
```

ServerBusinessHandler 对父类抽象方法 doSomethingElse(.)的重载实现了服务端的会话流程。当第一次文件接收完成之后连续两次发送文件，见第 15 行和第 17 行；当第二次文件接收完成之后关闭通道结束会话，见第 21 行。

类 ClientBusinessHandler 参见代码 4-21。

代码 4-21：类 ClientBusinessHandler

```
1  public class ClientBusinessHandler extends SaveFileHandler {
2      public ClientBusinessHandler(String savedFilePath) {
```

```
3           super(savedFilePath);
4       }
5       @Override
6       public void channelActive(ChannelHandlerContext ctx) throws Exception {
7           System.out.println("客户端建立连接,开始会话……");
8           String fPath = "D:\\ClientFiles\\abc.mp4";
9           ctx.channel().write(new SendCmd(fPath));
10      }
11      @Override
12      protected void doSomethingElse(ChannelHandlerContext ctx) {
13          //这里只需进行文件的发送
14          //文件的接收在父类的 channelRead(.)中完成
15          if(this.recvFileTimes == 2) {
16              String fPath = this.savedFileFolder + "abc.docx";
17              //因为要在下列的"写"之后关闭通道,而写操作是异步的
18              //所以关闭通道及其结束提示必须放在 Listener 里去做
19              //SendCmd(.)第二个参数为 true 指示发送成功之后关闭通道
20              ctx.channel().write(new SendCmd(fPath,true));
21          }
22      }
23 }
```

ClientBusinessHandler 对父类抽象方法 doSomethingElse(.)的重载完善了客户端的会话流程,当第二次文件接收完成时发送文件,见第 20 行。

服务端启动类参见代码 4-22。

代码 4-22:类 Server3

```
1  public class Server3 extends Server{
2      public static void main(String[] args) {
3          Server3 server = new Server3();
4          server.service();
5      }
6      public Server3(){
7          super(8001);
8      }
9      @Override
10     void addHandlers(SocketChannel ch) {
11         ch.pipeline().addLast(
                   new LengthFieldBasedFrameDecoder(Integer.MAX_VALUE, 0,4,0,4));
12         ch.pipeline().addLast(new ServerBusinessHandler("D:\\ServerFiles\\"));
13         ch.pipeline().addLast(new AttachHeaderHandler());
```

```
14            ch.pipeline().addLast(new SegFileHandler());
15        }
16 }
```

参照如图 4-10 所示的服务端,在重载的 addHandlers(.)里添加 ChannelHandler。

客户端启动类参见代码 4-23。

代码 4-23: 类 Client3

```
1  public class Client3 extends Client{
2      public static void main(String[] args) {
3          ExecutorService fixPool = Executors.newCachedThreadPool();
4          for (int i = 0; i < 1; i++) {
5              fixPool.execute(
6                      () -> {
7                          try {
8                              Client3 client = new Client3();
9                              client.communicate();
10                         } catch (Exception ex) {
11                             ex.printStackTrace();
12                         }
13                     }
14             );
15         }
16         fixPool.shutdown();
17     }
18     public Client3(){
19         super("127.0.0.1", 8001);
20     }
21     @Override
22     void addHandlers(SocketChannel ch) {
23         ch.pipeline().addLast(
                   new LengthFieldBasedFrameDecoder(Integer.MAX_VALUE, 0,4,0,4));
24         ch.pipeline().addLast(new ClientBusinessHandler("D:\\ClientFiles\\"));
25         ch.pipeline().addLast(new AttachHeaderHandler());
26         ch.pipeline().addLast(new SegFileHandler());
27     }
28 }
```

参照如图 4-10 所示的客户端,在重载的 addHandlers(.)里添加 ChannelHandler。为了避免单机上多个客户端对文件的访问冲突,这里只设置 1 个客户端。

运行时,先确保要传送的文件存在,再启动附件源代码 netty 项目的 Server3,然后启动

netty 项目的 Client3。

服务端及客户端控制台将分别显示如下信息，略去了文件片段发送成功时的提示信息。

服务端控制台信息	客户端控制台信息
服务器开启服务端口 8001，等待连接请求……	客户端建立连接，开始会话……
服务端建立连接，开始会话……	开始发送文件：D:\ClientFiles\abc. mp4
开始接收文件：abc. mp4	文件发送完成！ D:\ClientFiles\abc. mp4
保存文件完成！ D:\ServerFiles\8-abc. mp4	开始接收文件：xyz. mp4
开始发送文件：D:\ServerFiles\xyz. mp4	保存文件完成！ D:\ClientFiles\4-xyz. mp4
开始发送文件：D:\ServerFiles\xyz. pdf	开始接收文件：xyz. pdf
文件发送完成！ D:\ServerFiles\xyz. mp4	保存文件完成！ D:\ClientFiles\1-xyz. pdf
文件发送完成！ D:\ServerFiles\xyz. pdf	开始发送文件：D:\ClientFiles\abc. docx
开始接收文件：abc. docx	文件发送完成！ D:\ClientFiles\abc. docx
保存文件完成！ D:\ServerFiles\3-abc. docx	会话结束，客户端关闭连接！
会话结束，服务端关闭连接！	

习题

编写 Netty 程序，完成如下功能：客户端连接远程服务端，之后由使用者发送命令对服务端的文件系统进行操作，命令的顺序并不固定。

dir-列出当前文件夹内的文件及文件夹；

cd <文件夹>-改变服务端的当前文件夹；

download <远端文件名>-下载服务端的文件至本地；

upload <本地文件名>-上传本地文件至服务端；

close-结束会话。

第 5 章 Jersey

REST(REpresentational State Transfer)中文翻译成表征性状态转移,指的是一组架构约束条件和原则,符合下述REST主要原则的架构方式称为RESTful。

(1) 对网络上所有的资源都有一个资源标识符。

(2) 对资源的操作不会改变标识符。

(3) 同一资源有多种表现形式(如xml、json)。

(4) 所有操作都是无状态的。

REST本身并没有创造新的技术、组件或服务,而RESTful的理念就是使用Web的现有特征和能力,更好地使用现有Web标准中的一些准则和约束。目前,HTTP是唯一与REST相关的实例,因此通常所说的REST就是通过HTTP实现的REST。

在客户端和服务端之间传送的是资源的表述。资源的表述包括数据和描述数据的元数据,例如,HTTP头部的"Content-Type"属性就是这样一个元数据属性。双方这样通过HTTP进行协商:客户端通过HTTP请求的头部Accept属性请求一种特定格式的表述,服务端则通过HTTP应答的头部Content-Type属性告诉客户端资源的表述形式。当服务器不支持所请求的表述格式时,它应该返回一个HTTP 406状态码,表示拒绝处理该请求。

JAX-RS是RESTful Web服务的Java API规范,是使用Java开发REST服务时的基本约定。目前JAX-RS已经升级至JAX-RS 2.0,也称为JSR339。官方文档网址是http://www.jcp.org/en/jsr/detail?id=339。

Jersey是JAX-RS规范的参考实现,是Java领域里最符合JAX-RS规范的REST服务开发框架,Jersey 2.x对应JAX-RS 2.0,其官网为https://eclipse-ee4j.github.io/jersey/。

5.1 概述

使用Jersey开发服务端时基本的开发步骤如下。

(1) 导入依赖。

(2) 创建实体类。

(3) 创建资源类,设置资源 URI。

(4) 注册资源类。

使用 Jersey 开发客户端时基本的开发步骤如下。

(1) 导入依赖。

(2) 创建实体类。

(3) 访问服务端。

类 Student 和类 Book 是随后案例中都会用到的实体类,这两个实体类在服务端与客户端的定义相同,这里作统一说明,见代码 5-1,源码中这两个类必须分两个文件保存。

代码 5-1:实体类 Student 和实体类 Book

```
1  @Data
2  @AllArgsConstructor
3  @NoArgsConstructor
4  @XmlRootElement(name = "Student")
5  public class Student {
6      Integer id;
7      String name;
8      Sex gender;
9      @JsonSerialize(using = LocalDateSerializer.class)
10     @JsonDeserialize(using = LocalDateDeserializer.class)
11     LocalDate birthday;
12     double gpa;
13     List<Book> reads;
14 }
15
16 @Data
17 @AllArgsConstructor
18 @NoArgsConstructor
19 @XmlRootElement(name = "Book")
20 public class Book {
21     String name;
22     String author;
23     Double price;
24     int year;
25     Booktype type;
26 }
```

第 4 行和第 19 行是实体类与 XML 格式相互转换所必需的标注;第 11 行的 birthday 成员变量的两个标注(第 9 行和第 10 行)是 LocalDate 类型值与 JSON 相互转换时必需的标注。

随后的案例将采用下列 URI 与功能的对应关系，如表 5-1 所示。这里 URI 未包含 Application URI。

表 5-1　URI 与其对应的操作功能

URI	操作功能
GET /stu	查询全部学生，应答的 HTTP 消息体为学生信息列表
GET /stu/2	查询学号为 2 的学生，应答的 HTTP 消息体为学生信息
GET /stu/query?uid=2	查询学号为 2 的学生，应答的 HTTP 消息体为学生信息
POST /stu	新增学生信息，请求时 HTTP 消息体为要新增的学生信息，应答的 HTTP 消息体为已新增的学生信息
PUT/stu	更新学生信息，请求时 HTTP 消息体为要更新的学生信息，应答的 HTTP 消息体为已更新的学生信息
DELETE /stu/2	删除学号为 2 的学生信息
GET /file/abc.pdf	从服务端下载远程文件 abc.pdf
POST /file/xyz.mp4	向服务端上传本地文件 xyz.mp4

5.2　案例 1：对象资源的操作

本案例对应源代码 JerseyServer5 项目及 JerseyClient5 项目。

5.2.1　服务端基本框架

(1) 导入依赖，参见代码 5-2。

代码 5-2：服务端的依赖

```
1  <dependencies>
2      <!-- Spring 的自动配置功能 -->
3      <dependency>
4          <groupId>org.springframework.boot</groupId>
5          <artifactId>spring-boot-autoconfigure</artifactId>
6          <version>2.2.6.RELEASE</version>
7      </dependency>
8      <!-- 提供 Jersey 框架，内置 Tomcat -->
9      <dependency>
10         <groupId>org.springframework.boot</groupId>
11         <artifactId>spring-boot-starter-jersey</artifactId>
12         <version>2.2.6.RELEASE</version>
13     </dependency>
14     <dependency>
15         <groupId>org.projectlombok</groupId>
```

```
16            <artifactId>lombok</artifactId>
17            <version>1.18.12</version>
18        </dependency>
19        <!-- 下面两个依赖用来支持 XML 数据格式 -->
20        <dependency>
21            <groupId>org.glassfish.jersey.media</groupId>
22            <artifactId>jersey-media-moxy</artifactId>
23            <version>2.30.1</version>
24        </dependency>
25        <dependency>
26            <groupId>org.glassfish.jersey.media</groupId>
27            <artifactId>jersey-media-jaxb</artifactId>
28            <version>2.30.1</version>
29        </dependency>
30        <!-- 支持文件上传和下载 -->
31        <dependency>
32            <groupId>org.glassfish.jersey.media</groupId>
33            <artifactId>jersey-media-multipart</artifactId>
34            <version>2.30.1</version>
35        </dependency>
36 </dependencies>
```

这里导入了 spring-boot-starter-jersey 这个依赖，它自动包含了很多辅助的依赖。

(2) 创建两个空的资源类，分别对应 Student 和文件的操作。

代码 5-3：类 StudentResource(空的)

```
1  @Component
2  @Path("stu")
3  public class StudentResource {
4      //用于自增学号
5      private static AtomicInteger idCounter = new AtomicInteger(1000);
6      //将从 JSON 和 XML 中随机选择数据格式来发送应答
7      private final static MediaType[] mediaTypes = new MediaType[]{
                                    MediaType.APPLICATION_XML_TYPE,
                                    MediaType.APPLICATION_JSON_TYPE};
8      private MediaType randomMediaType(){
9          return mediaTypes[new Random().nextInt(mediaTypes.length)];
10     }
11 }
```

之后将补充 StudentResource 类的方法。@Path 表明该资源类对应的路径。

代码 5-4：类 FileResource(空的)

```
1  @Component
2  @Path("file")
```

```
3  public class FileResource {
4      private final String folderOnServer = "D:\\ServerFiles\\";
5  }
```

之后将补充 FileResource 类的方法。@Path 表明该资源类对应的路径。

(3) 注册资源类。

在 JerseyConfig 的构造器中完成资源类的注册。

代码 5-5：类 JerseyConfig

```
1  @Component
2  @ApplicationPath("/jersey")
3  public class JerseyConfig extends ResourceConfig {
4      public JerseyConfig(){
5          register(StudentResource.class);
6          register(FileResource.class);
7          register(MoxyXmlFeature.class);
8          register(JacksonFeature.class);
9          register(MultiPartFeature.class);
10     }
11 }
```

第 5 行和第 6 行注册新建的两个空资源类,第 7 行注册 XML 解析器,第 8 行注册 JSON 解析器,第 9 行注册的类用于支持文件传输。第 2 行的标注指明应用路径。

(4) 服务启动类。

代码 5-6：类 Server

```
1  @SpringBootApplication
2  public class Server {
3      public static void main(String[] args) {
4          SpringApplication.run(Server.class, args);
5      }
6  }
```

5.2.2 客户端基本框架

(1) 导入客户端依赖,参见代码 5-7。

代码 5-7：客户端依赖

```
1  <dependencies>
2      <dependency>
3          <groupId>org.projectlombok</groupId>
4          <artifactId>lombok</artifactId>
```

```
5            <version>1.18.10</version>
6        </dependency>
7        <dependency>
8            <groupId>org.glassfish.jersey.core</groupId>
9            <artifactId>jersey-client</artifactId>
10           <version>2.30.1</version>
11       </dependency>
12       <!--
13           下面这三个依赖用来支持 JSON 数据格式
14           spring-boot-starter-jersey 已经包括下列这三个依赖
15           故服务端不需要导入这三个依赖
16       -->
17       <dependency>
18           <groupId>org.glassfish.jersey.inject</groupId>
19           <artifactId>jersey-hk2</artifactId>
20           <version>2.30.1</version>
21       </dependency>
22       <dependency>
23           <groupId>org.glassfish.jersey.media</groupId>
24           <artifactId>jersey-media-json-jackson</artifactId>
25           <version>2.30.1</version>
26       </dependency>
27       <dependency><!-- 这个用于解决 LocalDate 类型与 JSON 相互转换的问题 -->
28           <groupId>com.fasterxml.jackson.datatype</groupId>
29           <artifactId>jackson-datatype-jsr310</artifactId>
30           <version>2.11.2</version>
31       </dependency>
32       <!--
33       下面这两个依赖加上前面已经导入的 jersey-hk2
34       三个一起用来支持 XML 数据格式
35       -->
36       <dependency>
37           <groupId>org.glassfish.jersey.media</groupId>
38           <artifactId>jersey-media-moxy</artifactId>
39           <version>2.30.1</version>
40       </dependency>
41       <dependency>
42           <groupId>org.glassfish.jersey.media</groupId>
43           <artifactId>jersey-media-jaxb</artifactId>
44           <version>2.30.1</version>
45       </dependency>
46       <!-- 支持文件上传和下载 -->
47       <dependency>
48           <groupId>org.glassfish.jersey.media</groupId>
49           <artifactId>jersey-media-multipart</artifactId>
```

```
50          <version>2.30.1</version>
51      </dependency>
52    /dependencies>
```

(2) 客户端启动类。

代码 5-8：类 Client

```
1  public class Client {
2      public static void main(String[] args) {
3          Client client = new Client();
4          client.doWork();
5      }
6      private final String folderOnClient = "D:\\ClientFiles\\";
7      //将随机选择数据格式来发送服务请求
8      private final static MediaType[] mediaTypes = new MediaType[]{
                                   MediaType.APPLICATION_XML_TYPE,
                                   MediaType.APPLICATION_JSON_TYPE};
9      private MediaType randomMediaType(){
10         return mediaTypes[new Random().nextInt(mediaTypes.length)];
11     }
12     private WebTarget target;
13     public Client(){
14         ClientConfig clientConfig = new ClientConfig();
15         clientConfig.register(MoxyXmlFeature.class);
16         clientConfig.register(JacksonFeature.class);
17         clientConfig.register(MultiPartFeature.class);
18         javax.ws.rs.client.Client rsClient = ClientBuilder.newClient(clientConfig);
19         target = rsClient.target("http://localhost:8080/jersey/");
20     }
21     public void doWork(){     }
22 }
```

第 15 行注册 XML 解码器，第 16 行注册 JSON 解码器，第 17 行注册的类用于支持文件传输，第 19 行的 target(.)方法的参数指明了服务端应用路径。第 21 行的 doWork()方法有待补充。

5.2.3 逐项添加 URI 功能

在开始添加 URI 之前，先了解传输格式的设置，参见图 5-1。

当客户端发出 HTTP 请求时，需要确定请求头部 Accept 属性和 Content-Type 属性的值，这可以通过代码中 target.request(.)和 Entity.entity(.)的参数来指定；服务端应答时，需要确定应答头部 Content-Type 属性的值，这可以通过代码中 Response.type(.)的参数来

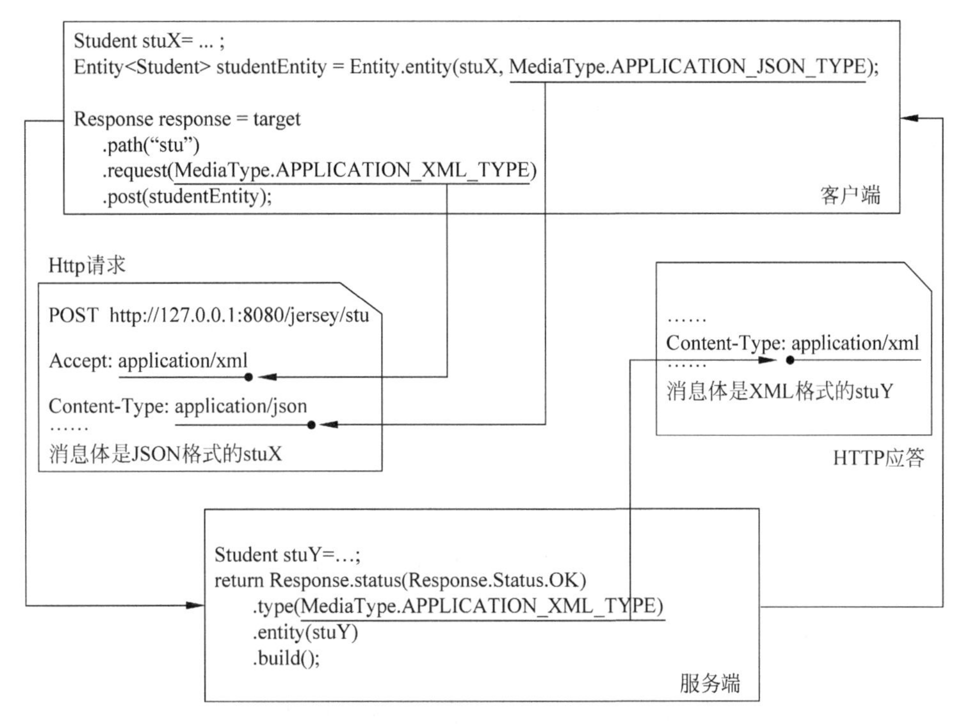

图 5-1 请求及应答时传输格式的设置

指定。应答消息体的默认格式与请求时 Accept 属性指定的格式保持一致，但也可以采用 Accept 属性未指定的格式，只需要客户端注册了对应的格式解析器即可。服务端应答时，消息体格式的选择范围仅限于此 URI 方法的@Produces 标注所设定的格式列表；服务端接收请求时允许的请求消息体格式仅限于从此 URI 方法的@Consumes 标注所设定的格式列表中选择。

1. GET /stu

服务端方法见代码 5-9，此方法添加于 StudentResource 类。

代码 5-9：服务端函数 getAll()

```
1  @GET
2  @Produces({MediaType.APPLICATION_XML, MediaType.APPLICATION_JSON})
3  public List<Student> getAll(){
4      List<Student> students = StuRepository.getStudents();
5      return students;
6  }
```

这种直接返回实体对象的应答，其消息体将使用请求时 Accept 属性值指定的格式，且应答状态码为 OK。

客户端对应的请求函数见代码 5-10。

代码 5-10：客户端函数 getAllStudents()

```
1  private List<Student> getAllStudents(){
2      MediaType mediaType = randomMediaType();
3      Response response = target
4              .path("stu")
5              .request(mediaType)
6              .get();
7      GenericType<List<Student>> genericType = new GenericType<>(){ };
8      List<Student> students = response.readEntity(genericType);
9      return students;
10 }
```

第 2 行调用的函数 randomMediaType()，其定义见代码 5-8 的第 9 行。

如果服务端应答的是列表，客户端必须使用 GenericType 泛型来获取列表对象，见第 7 行与第 8 行。如果服务端函数返回值为 null，则客户端从 response 解析出来的对象也是 null。

2. GET /stu/2

服务端方法见代码 5-11，此方法添加于 StudentResource 类。

代码 5-11：服务端函数 getOne(.)

```
1  @GET
2  @Path("{id}")
3  @Produces({MediaType.APPLICATION_XML, MediaType.APPLICATION_JSON})
4  public Student getOne(@PathParam("id") Integer id) {
5      Student student = StuRepository.getStudent(id);
6      return student;
7  }
```

@PathParam 用来注入路径参数。

客户端对应的请求函数见代码 5-12。

代码 5-12：客户端函数 getStudentById(.)

```
1  private Student getStudentById(int sId){
2      MediaType mediaType = randomMediaType();
3      Response response = target
4              .path("stu")
5              .path(String.valueOf(sId))
6              .request(mediaType)
7              .get();
8      Student st = response.readEntity(Student.class);
9      return  st;
10 }
```

第 5 行用.path(.)来添加路径参数。

3. GET /stu/query?uid=2

服务端方法见代码 5-13，此方法添加于 StudentResource 类。

代码 5-13：服务端函数 getOneByQuery(.)

```
1  @GET
2  @Path("query")
3  @Produces({MediaType.APPLICATION_XML, MediaType.APPLICATION_JSON})
4  public Student getOneByQuery(@QueryParam("uid") Integer id) {
5      Student student = StuRepository.getStudent(id);
6      return student;
7  }
```

@QueryParam 用来注入查询参数。

客户端对应的请求函数见代码 5-14。

代码 5-14：客户端函数 queryStudentById(.)

```
1  private Student queryStudentById(int sId){
2      MediaType mediaType = randomMediaType();
3      Response response = target
4              .path("stu")
5              .path("query")
6              .queryParam("uid", sId)
7              .request(mediaType)
8              .get();
9      Student st = response.readEntity(Student.class);
10     return st;
11 }
```

第 6 行用.queryParam(.)添加查询参数，可以多次调用.queryParam(.)以添加多个查询参数。

4. POST /stu

服务端方法见代码 5-15，此方法添加于 StudentResource 类。

代码 5-15：服务端函数 createStudent(.)

```
1  @POST
2  @Consumes({MediaType.APPLICATION_XML, MediaType.APPLICATION_JSON})
3  @Produces({MediaType.APPLICATION_XML, MediaType.APPLICATION_JSON})
4  public Student createStudent(Student student){
5      student.setId(idCounter.incrementAndGet());
6      StuRepository.insertStudent(student);
```

```
7       return student;
8   }
```

第 7 行若改成“return Response. ok(student). build();”，则需要修改此函数返回值类型为 Response。

客户端对应的请求函数见代码 5-16。

代码 5-16：客户端函数 addStudent(.)

```
1  private Student addStudent(){
2       Student student = newRandomStudent();
3       MediaType mediaType = randomMediaType();
4       Entity<Student> studentEntity = Entity.entity(student, mediaType);
5       mediaType = randomMediaType();
6       Response response = target
7               .path("stu")
8               .request(mediaType)
9               .post(studentEntity);
10      Student stu = response.readEntity(Student.class);
11      return stu;
12 }
```

如果服务端函数返回值为 null，则客户端从 response 解析出来的对象也是 null。

5. PUT/stu

服务端方法见代码 5-17，此方法添加于 StudentResource 类。

代码 5-17：服务端函数 modifyStudent(.)

```
1  @PUT
2  @Consumes({MediaType.APPLICATION_XML, MediaType.APPLICATION_JSON})
3  @Produces({MediaType.APPLICATION_XML, MediaType.APPLICATION_JSON})
4  public Response modifyStudent(Student student){
5       //模拟对学生信息的更新
6       Student stu = StuRepository.updateStudent(student);
7       if (stu!= null){
8           return Response
9                   .status(Response.Status.OK)
10                  .type(randomMediaType())
11                  .entity(stu)
12                  .build();
13      }else{
14          return Response
15                  .status(Response.Status.NOT_MODIFIED)
16                  .build();
```

```
17      }
18 }
```

因为可能返回不同的状态码，故此函数使用 Response 作为返回类型。第 10 行若不调用 .type(.)，则自动使用请求时 Accept 属性值。第 15 行应答的状态码为 NOT_MODIFIED，HTTP 规定状态码为 NOT_MODIFIED 时不会产生消息体，因此这里只需要提供状态码，即使调用了 .entity(.)及 .type(.)也不会产生应答的消息体。

客户端对应的请求函数见代码 5-18。

代码 5-18：客户端函数 updateStudent(.)

```
1  private Student updateStudent(int id){
2      //模拟出学生的更新信息
3      Student st = newRandomStudent();
4      st.setId(id);
5      MediaType mediaType = randomMediaType();
6      Entity<Student> studentEntity = Entity.entity(st, mediaType);
7      mediaType = randomMediaType();
8      Response response = target
9              .path("stu")
10             .request(mediaType)
11             .put(studentEntity);
12     Student stu = null;
13     //根据应答的状态码作不同处理
14     if (response.getStatus() == Response.Status.NOT_MODIFIED.getStatusCode()){
15         System.out.println("修改失败，编号：" + id);
16     }else{
17         stu = response.readEntity(Student.class);
18         System.out.println("修改成功，编号：" + stu.getId());
19     }
20     return stu;
21 }
```

第 14 行也可以通过检查 response.hasEntity()返回值是否为 true 来知晓应答是否包含消息体，以避免在无消息体时调用 response.readEntity(.)造成异常。

6. DELETE /stu/2

服务端方法见代码 5-19，此方法添加于 StudentResource 类。

代码 5-19：服务端函数 eraseStudent(.)

```
1  @DELETE
2  @Path("{id}")
3  public Response eraseStudent(@PathParam("id") Integer id){
```

```
4        if (StuRepository.deleteStudent(id)){
5            return Response.ok(id.toString()).build();
6        }
7        return Response.status(Response.Status.NOT_FOUND)
8                .entity(id.toString())
9                .type(randomMediaType())
10               .build();
11 }
```

应答时除了返回状态码，还把学号作为消息体字符串返回。

客户端对应的请求函数见代码 5-20。

代码 5-20：客户端函数 deleteStudent(.)

```
1  private boolean deleteStudent(int id){
2      MediaType mediaType = randomMediaType();
3      Response response = target
4              .path("stu")
5              .path(String.valueOf(id))
6              .request(mediaType)
7              .delete();
8      //根据应答的状态码作不同处理
9      if (response.getStatus() == 200){
10         System.out.println("删除成功，编号：" + response.readEntity(String.class));
11     }else{
12         System.out.println("删除失败，编号：" + response.readEntity(String.class));
13     }
14     return (response.getStatus() == 200);
15 }
```

第 10 行和第 12 行读取消息体字符串。

7. GET /file/abc.pdf

服务端方法见代码 5-21，此方法添加于 FileResource 类。

代码 5-21：服务端函数 download(.)

```
1  @GET
2  @Path("/{name}")
3  public Response download(@PathParam("name") String fileName)
       throws IOException {
4      File f = new File(this.folderOnServer + fileName);
5      if (!f.exists()) {
6          return Response.status(Response.Status.NOT_FOUND).build();
7      } else {
8          return Response
```

```
9                   .ok(f)
10                  .header("Content-disposition", "attachment;filename=" + fileName)
11                  .header("Cache-Control", "no-cache")
12                  .build();
13      }
14 }
```

这里不需要另行设定.type(.),而是采用请求时头部 Accept 属性指定的数据格式。

客户端对应的请求函数见代码 5-22。

代码 5-22：客户端函数 download(.)

```
1  private void download(String remoteFileName){
2      Response response = target
3              .path("file")
4              .path(remoteFileName)
5              .request(MediaType.APPLICATION_OCTET_STREAM_TYPE)
6              .get();
7      if (response.getStatus() == Response.Status.NOT_FOUND.getStatusCode()){
8          System.out.println("远程文件不存在：" + remoteFileName);
9      }
10     if (response.getStatus() == Response.Status.OK.getStatusCode()) {
11         //下载至本地时文件名自动增加 UUID 以避免和已有文件重名
12         String localFilePath = folderOnClient +
                                   UUID.randomUUID().toString() + "-" + remoteFileName;
13         File localFile = new File(localFilePath);
14         byte[] bytes = response.readEntity(byte[].class);
15         try {
16             OutputStream fos = new FileOutputStream(localFile);
17             fos.write(bytes);
18             fos.flush();
19             fos.close();
20             System.out.println("保存文件成功：" + localFilePath);
21         } catch (Exception e) {
22             e.printStackTrace();
23         }
24     }
25 }
```

第 5 行是下载文件时专有的请求数据格式,作为请求头部属性 Accept 的值。

8. POST /file/xyz.mp4

服务端方法见代码 5-23,此方法添加于 FileResource 类。

代码 5-23：服务端函数 upload()

```
1  @POST
2  @Consumes(MediaType.MULTIPART_FORM_DATA)
3  public Response upload(@FormDataParam("file") FormDataBodyPart bp) {
4      FormDataContentDisposition disposition = bp.getFormDataContentDisposition();
5      String fileName = disposition.getFileName();
6      //上传至服务端时文件名自动增加 UUID 以避免和已有文件重名
7      File file = new File(this.folderOnServer +
                       UUID.randomUUID().toString() + "-" + fileName);
8      try {
9          InputStream is = bp.getValueAs(InputStream.class);
10         OutputStream fos = new FileOutputStream(file);
11         byte[] buffer = new byte[1024 * 1024];
12         int len = 0;
13         while( (len = is.read(buffer)) != -1 ){
14             fos.write(buffer, 0, len);
15             fos.flush();
16         }
17         fos.close();
18     } catch (IOException e) {
19         e.printStackTrace();
20         return Response.notModified().build();
21     }
22     return Response.ok().build();
23 }
```

客户端对应的请求函数见代码 5-24。

代码 5-24：客户端函数 upload()

```
1  private void upload(String localFileName){
2      String localFilePath = folderOnClient + localFileName;
3      File localFile = new File(localFilePath);
4      if (!localFile.exists()){
5          System.out.println("本地文件不存在：" + localFilePath);
6          return;
7      }
8      FormDataMultiPart part = new FormDataMultiPart();
9      //"file"是控件命名，必须与服务端的@FormDataParam("file")一致
10     part.bodyPart(new FileDataBodyPart("file",localFile));
11     Response response = target
12             .path("file")
13             .request(randomMediaType())
14             .post(Entity.entity(part, MediaType.MULTIPART_FORM_DATA_TYPE));
15     if (response.getStatus() == Response.Status.OK.getStatusCode()){
16         System.out.println("上传文件成功：" + localFilePath);
```

```
17      }else{
18          System.out.println("上传文件失败：" + localFilePath);
19      }
20 }
```

最后补充客户端 doWork()函数，如代码 5-25 所示。

代码 5-25：客户端函数 doWork()

```
1  public void doWork(){
2      getAllStudents();
3      getStudentById(102);
4      queryStudentById(103);
5      addStudent();
6      updateStudent(104);
7      updateStudent(1050);
8      deleteStudent(103);
9      deleteStudent(1060);
10     download("xyz.pdf");
11     download("xyz.mp4");
12     download("12345.pdf");
13     upload("abc.docx");
14     upload("abc.txt");
15     upload("12345.mp4");
16 }
```

运行时先启动源码中项目 JerseyServer5 的 Server 类，再启动源码中项目 JerseyClient5 的 Client 类。

5.3　案例 2：异步请求与异步应答

本案例对应源代码 JerseyServer6 项目及 JerseyClient6 项目。

客户端的异步请求与服务端的异步应答并不要求双方同时使用。

5.3.1　服务端基本框架

除了服务启动类，其他各项与 5.2.1 节相同。

代码 5-26：类 Server

```
1  @SpringBootApplication
2  public class Server {
3      public static final ExecutorService pool = Executors.newCachedThreadPool();
```

```
4       public static void main(String[ ] args) {
5           SpringApplication.run(Server.class, args);
6       }
7   }
```

启动类 Server 的第 3 行增加一个线程池 pool，用于支持服务端的异步处理。

5.3.2 客户端基本框架

与 5.2.2 节相同。

5.3.3 逐项添加 URI 功能

1. GET /stu

服务端方法见代码 5-27，此方法添加于 StudentResource 类。

代码 5-27：服务端函数 getAll_Asyn()

```
1  @GET
2  @Produces({MediaType.APPLICATION_XML, MediaType.APPLICATION_JSON})
3  public void getAll_Asyn(@Suspended final AsyncResponse asyncResponse){
4      Server.pool.submit(
5              () ->{
6                  asyncResponse.resume(getAll());
7              }
8      );
9  }
10 private List<Student> getAll(){
11     List<Student> students = StuRepository.getStudents();
12     return students;
13 }
```

第 10 行的 getAll()函数与代码 5-9 的函数相同，无须加上任何标注。服务端异步应答就是把同步应答的功能函数放到线程池里执行，并通过 asyncResponse.resume(.)把同步应答功能函数的返回值以异步的方式答复给客户端。

客户端对应的请求函数见代码 5-28。

代码 5-28：客户端函数 getAllStudents_Asyn()

```
1  private List<Student> getAllStudents_Asyn(){
2      final AsyncInvoker async = target
3              .path("stu")
```

```
4              .request()      //头部属性 Accept 对应的参数默认值为 json
5              .async();
6      final Future<Response> responseFuture = async
7              .get(
8                      new InvocationCallback<Response>(){
9                          @Override
10                         public void completed(Response response) {
11                             //此处可以记录日志等
12                         }
13                         @Override
14                         public void failed(Throwable throwable) {
15                             throwable.printStackTrace();
16                         }
17                     }
18             );
19     GenericType<List<Student>> genericType = new GenericType<>(){ };
20     List<Student> students = null;
21     try{
22         Response response = responseFuture.get();
23         students = response.readEntity(genericType);
24     }catch (Exception e){
25         e.printStackTrace();
26     }
27     return students;
28 }
```

在第4行的.request()之后增加调用.async()；获得一个AsyncInvoker对象，然后执行这个AsyncInvoker对象的get(.)方法向服务端发出GET请求，在第10行的回调函数里可以得到服务端的应答，也可以如第22行所示调用responseFuture.get()等待直到获得服务端的应答，注意，这个responseFuture.get()与HTTP的GET方法没有任何关系。虽然服务端功能函数的返回值为List<Student>类型，但是客户端还是使用泛型Future<Response>，见第6行；另外，第23行的response.readEntity(.)只能调用一次，重复调用会发生异常，因此如果在回调函数里调用了response.readEntity(.)，则在主线程里不能再调用response.readEntity(.)。

2. GET /stu/2

服务端方法见代码5-29，此方法添加于StudentResource类。

代码5-29：服务端函数getOne_Asyn()

```
1  @GET
2  @Path("{id}")
```

```
3  @Produces({MediaType.APPLICATION_XML, MediaType.APPLICATION_JSON})
4  public void getOne_Asyn(
                        @Suspended final AsyncResponse asyncResponse,
                        @PathParam("id") Integer id) {
5      Server.pool.submit(
6              () ->{
7                  asyncResponse.resume(getOne(id));
8              }
9      );
10 }
11 private Student getOne(Integer id) {
12     Student student = StuRepository.getStudent(id);
13     return student;
14 }
```

第11行的getOne(.)函数与代码5-11的函数相同，无须加上任何标注。服务端异步应答就是把同步应答的功能函数放到线程池里执行，并通过asyncResponse.resume(.)把同步应答功能函数的返回值以异步的方式答复给客户端。

客户端对应的请求函数见代码5-30。

代码5-30：客户端函数getStudentById_Asyn(.)

```
1  private Student getStudentById_Asyn(int sId){
2      MediaType mediaType = randomMediaType();
3      final AsyncInvoker async = target
4              .path("stu")
5              .path(String.valueOf(sId))
6              .request(mediaType)
7              .async();
8      final Future<Response> responseFuture = async
9              .get(
10                     new InvocationCallback<Response>() {
11                         @Override
12                         public void completed(Response response) {
13                             //此处可以记录日志等
14                         }
15                         @Override
16                         public void failed(Throwable throwable) {
17                             throwable.printStackTrace();
18                         }
19                     }
20             );
21     Student st = null;
22     try {
23         Response response = responseFuture.get();
```

```
24            st = response.readEntity(Student.class);
25        }catch (Exception e){
26            e.printStackTrace();
27        }
28        return st;
29 }
```

在第 6 行的.request(.)之后增加调用.async()；获得一个 AsyncInvoker 对象，然后执行这个 AsyncInvoker 对象的 get(.)方法向服务端发出 GET 请求，在第 12 行的回调函数里可以得到服务端的应答，也可以如第 23 行所示调用 responseFuture.get()等待直到获得服务端的应答，注意，这个 responseFuture.get()与 HTTP 的 GET 方法没有任何关系。虽然服务端功能函数的返回值为 Student 类型，但是客户端还是使用泛型 Future<Response>，见第 8 行；另外，第 24 行的 response.readEntity(.)只能调用一次，重复调用会发生异常，因此如果在回调函数里调用了 response.readEntity(.)，则在主线程里不能再调用 response.readEntity(.)。

3. GET /stu/query?uid=2

服务端方法见代码 5-31，此方法添加于 StudentResource 类。

代码 5-31：服务端函数 getOneByQuery_Asyn(.)

```
1  @GET
2  @Path("query")
3  @Produces({MediaType.APPLICATION_XML, MediaType.APPLICATION_JSON})
4  public void getOneByQuery_Asyn(
                          @Suspended final AsyncResponse asyncResponse,
                          @QueryParam("uid") Integer id) {
5      Server.pool.submit(
6              () ->{
7                  asyncResponse.resume(getOneByQuery(id));
8              }
9      );
10 }
11 private Student getOneByQuery(Integer id) {
12     Student student = StuRepository.getStudent(id);
13     return student;
14 }
```

第 11 行的 getOneByQuery(.)函数与代码 5-13 的函数相同，无须加上任何标注。服务端异步应答就是把同步应答的功能函数放到线程池里执行，并通过 asyncResponse.resume(.)把同步应答功能函数的返回值以异步的方式答复给客户端。

客户端对应的请求函数见代码 5-32。

代码 5-32：客户端函数 queryStudentById_Asyn(.)

```
1  private Student queryStudentById_Asyn(int sId){
2      MediaType mediaType = randomMediaType();
3      final AsyncInvoker async = target
4              .path("stu")
5              .path("query")
6              .queryParam("uid", sId)
7              .request(mediaType)
8              .async();
9      final Future<Response> responseFuture = async
10             .get(
11                     new InvocationCallback<Response>() {
12                         @Override
13                         public void completed(Response response) {
14                             //此处可以记录日志等
15                         }
16                         @Override
17                         public void failed(Throwable throwable) {
18                             throwable.printStackTrace();
19                         }
20                     }
21             );
22     Student st = null;
23     try {
24         Response response = responseFuture.get();
25         st = response.readEntity(Student.class);
26     }catch (Exception e){
27         e.printStackTrace();
28     }
29     return st;
30 }
```

在第 7 行的.request(.)之后增加调用.async()；获得一个 AsyncInvoker 对象，然后执行这个 AsyncInvoker 对象的 get(.)方法向服务端发出 GET 请求，在第 13 行的回调函数里可以得到服务端的应答，也可以如第 24 行所示调用 responseFuture.get()等待直到获得服务端的应答，注意，这个 responseFuture.get()与 HTTP 的 GET 方法没有任何关系。虽然服务端功能函数的返回值为 Student 类型，但是客户端还是可以使用泛型 Future<Response>，见第 9 行；另外，第 25 行的 response.readEntity(.)只能调用一次，重复调用会发生异常，因此如果在回调函数里调用了 response.readEntity(.)，则在主线程里不能再调用 response.readEntity(.)。

4. POST /stu

服务端方法见代码 5-33,此方法添加于 StudentResource 类。

代码 5-33：服务端函数 createStudent_Asyn(.)

```
1  @POST
2  @Consumes({MediaType.APPLICATION_XML, MediaType.APPLICATION_JSON})
3  @Produces({MediaType.APPLICATION_XML, MediaType.APPLICATION_JSON})
4  public void createStudent_Asyn(
                               @Suspended final AsyncResponse asyncResponse,
                               Student student){
5      Server.pool.submit(
6              () ->{
7                  asyncResponse.resume(createStudent(student));
8              }
9      );
10 }
11 private Student createStudent(Student student){
12     student.setId(idCounter.incrementAndGet());
13     if (StuRepository.insertStudent(student)){
14         return student;
15     }
16     return null;
17 }
```

第 11 行的 createStudent(.)函数与代码 5-15 的函数功能一致,无须加上任何标注。服务端异步应答就是把同步应答的功能函数放到线程池里执行,并通过 asyncResponse.resume(.)把同步应答功能函数的返回值以异步的方式答复给客户端。

客户端对应的请求函数见代码 5-34。

代码 5-34：客户端函数 addStudent_Asyn(.)

```
1  private Student addStudent_Asyn(){
2      Student student = newRandomStudent();
3      MediaType mediaType = randomMediaType();
4      Entity<Student> studentEntity = Entity.entity(student, mediaType);
5      mediaType = randomMediaType();
6      final AsyncInvoker async = target
7              .path("stu")
8              .request(mediaType)
9              .async();
10     //因为服务端功能函数的返回值为 Student 类型,
11     //所以客户端可以直接使用泛型 Future<Student>
12     final Future<Student> studentFuture = async
```

```
13              .post(
14                      studentEntity,
15                      new InvocationCallback<Student>() {
16                          @Override
17                          public void completed(Student stu) {
18                              if (stu!= null) {
19                                  System.out.println("添加成功,编号:" + stu.getId());
20                              }else{
21                                  System.out.println("添加用户失败!");
22                              }
23                          }
24                          @Override
25                          public void failed(Throwable throwable) {
26                              throwable.printStackTrace();
27                          }
28                      }
29              );
30      Student stu = null;
31      try{
32          //若添加不成功,得到的 stu 是 null
33          stu = studentFuture.get();
34      }catch (Exception e){
35          e.printStackTrace();
36      }
37      return stu;
38 }
```

在第 8 行的.request(.)之后增加调用.async();获得一个 AsyncInvoker 对象,然后执行这个 AsyncInvoker 对象的 post(.)方法向服务端发出 POST 请求,在第 17 行的回调函数里可以得到服务端的应答,也可以如第 33 行所示调用 studentFuture.get()等待直到获得服务端的应答。因为服务端功能函数的返回值为 Student 类型,所以客户端可以直接使用泛型 Future<Student>,见第 12 行;这种直接用 Future<Student>而不是用 Future<Response>的方式使得在回调函数和主线程都可以使用应答的 Student 对象,避免了因使用 Future<Response>泛型导致只能调用一次 response.readEntity(.)的不足。

5. PUT/stu

服务端方法见代码 5-35,此方法添加于 StudentResource 类。

代码 5-35:服务端函数 modifyStudent_Asyn(.)

```
1  @PUT
2  @Consumes({MediaType.APPLICATION_XML, MediaType.APPLICATION_JSON})
3  @Produces({MediaType.APPLICATION_XML, MediaType.APPLICATION_JSON})
```

```
4  public void modifyStudent_Asyn(
                          @Suspended final AsyncResponse asyncResponse,
                          Student student){
5      Server.pool.submit(
6              () ->{
7                  asyncResponse.resume(modifyStudent(student));
8              }
9      );
10 }
11 private Response modifyStudent(Student student){
12     fakeLongTimeOperation();     //模拟一个很耗时的操作
13     Student stu = StuRepository.updateStudent(student);
14     if (stu!= null){
15         return Response
16                 .status(Response.Status.OK)
17                 .type(randomMediaType())
18                 .entity(stu)
19                 .build();
20     }else{
21         return Response
22                 .status(Response.Status.NOT_MODIFIED)
23                 .build();
24     }
25 }
```

第 11 行的 modifyStudent(.)函数与代码 5-17 的函数相同,无须加上任何标注。服务端异步应答就是把同步应答的功能函数放到线程池里执行,并通过 asyncResponse.resume(.)把同步应答功能函数的返回值以异步的方式答复给客户端。

客户端对应的请求函数见代码 5-36。

代码 5-36:客户端函数 updateStudent_Asyn(.)

```
1  private Student updateStudent_Asyn(int id){
2      Student st = newRandomStudent();
3      st.setId(id);
4      MediaType mediaType = randomMediaType();
5      Entity<Student> studentEntity = Entity.entity(st, mediaType);
6      mediaType = randomMediaType();
7      final AsyncInvoker async = target
8              .path("stu")
9              .request(mediaType)
10             .async();
11     //因为服务端功能函数的返回值为 Response 类型,
12     //所以客户端使用泛型 Future<Response>
13     final Future<Response> responseFuture = async
```

```
14              .put(
15                      studentEntity,
16                      new InvocationCallback<Response>() {
17                          @Override
18                          public void completed(Response response) {
19                              //根据响应的状态码作不同处理
20                              if (response.getStatus() ==
                                    Response.Status.NOT_MODIFIED.getStatusCode()){
21                                  System.out.println("修改失败,编号: " + id);
22                              }else{
23                                  System.out.println("修改成功,编号: " + id);
24                              }
25                          }
26                          @Override
27                          public void failed(Throwable throwable) {
28                              throwable.printStackTrace();
29                          }
30                      }
31              );
32      Response response = null;
33      Student stu = null;
34      try{
35          response = responseFuture.get();
36      }catch (Exception e){
37          e.printStackTrace();
38      }
39      if (response.getStatus() == Response.Status.OK.getStatusCode()){
40          stu = response.readEntity(Student.class);
41      }
42      return stu;
43 }
```

在第9行的.request(.)之后增加调用.async();获得一个AsyncInvoker对象,然后执行这个AsyncInvoker对象的put(.)方法向服务端发出PUT请求,在第18行的回调函数里可以得到服务端的应答,也可以如第35行所示调用responseFuture.get()等待直到获得服务端的应答。因为服务端功能函数的返回值为Response类型,所以客户端使用泛型Future<Response>,见第13行;另外,第40行的response.readEntity(.)只能调用一次,重复调用会发生异常,因此如果在回调函数里调用了response.readEntity(.),则在主线程的第40行不能再调用response.readEntity(.)。

6. DELETE /stu/2

服务端方法见代码5-37,此方法添加于StudentResource类。

代码 5-37：服务端函数 eraseStudent_Asyn(.)

```
1  @DELETE
2  @Path("{id}")
3  public void eraseStudent_Asyn(
                               @Suspended final AsyncResponse asyncResponse,
                               @PathParam("id") Integer id) {
4      Server.pool.submit(
5              () ->{
6                  asyncResponse.resume(eraseStudent(id));
7              }
8      );
9  }
10 private Response eraseStudent(Integer id){
11     if (StuRepository.deleteStudent(id)){
12         return Response.ok(id.toString()).build();
13     }
14     return Response.status(Response.Status.NOT_FOUND)
15             .entity(id.toString())
16             .type(randomMediaType())
17             .build();
18 }
```

第 10 行的 eraseStudent(.)函数与代码 5-19 的函数相同，无须加上任何标注。服务端异步应答就是把同步应答的功能函数放到线程池里执行，并通过 asyncResponse.resume(.)把同步应答功能函数的返回值以异步的方式答复给客户端。

客户端对应的请求函数见代码 5-38。

代码 5-38：客户端函数 deleteStudent_Asyn(.)

```
1  private boolean deleteStudent_Asyn(int id){
2      MediaType mediaType = randomMediaType();
3      final AsyncInvoker async = target
4              .path("stu")
5              .path(String.valueOf(id))
6              .request(mediaType)
7              .async();
8      final Future<Response> responseFuture = async
9              .delete(
10                     new InvocationCallback<Response>() {
11                         @Override
12                         public void completed(Response response) {
13                             //根据响应的状态码作不同处理
14                             if (response.getStatus() == 200){
15                                 System.out.println("删除成功，编号：" +
                                                 response.readEntity(String.class));
```

```
16                                  }else{
17                                      System.out.println("删除失败,编号:" +
                                                    response.readEntity(String.class));
18                                  }
19                              }
20                              @Override
21                              public void failed(Throwable throwable) {
22                                  throwable.printStackTrace();
23                              }
24                          }
25              );
26      Response response = null;
27      try{
28          response = responseFuture.get();
29      }catch (Exception e){
30          e.printStackTrace();
31      }
32      return (response.getStatus() == 200);
33 }
```

在第 6 行的.request(.)之后增加调用.async();获得一个 AsyncInvoker 对象,然后执行这个 AsyncInvoker 对象的 delete(.)方法向服务端发出 DELETE 请求,在第 12 行的回调函数里可以得到服务端的应答,也可以如第 28 行所示调用 responseFuture.get()等待直到获得服务端的应答。因为服务端功能函数的返回值为 Response 类型,所以客户端使用泛型 Future<Response>,见第 8 行;另外,response.readEntity(.)只能调用一次,重复调用会发生异常,因此在回调函数的第 15 行和第 17 行调用了 response.readEntity(.),则在主线程里不能再调用 response.readEntity(.)。

7. GET /file/abc.pdf

服务端方法见代码 5-39,此方法添加于 FileResource 类。

代码 5-39:服务端函数 download_Asyn(.)

```
1  @GET
2  @Path("/{name}")
3  public void download_Asyn(
                        @Suspended final AsyncResponse asyncResponse,
                        @PathParam("name") String fileName){
4      Server.pool.submit(
5              () ->{
6                  try {
7                      asyncResponse.resume(download(fileName));
8                  }catch (Exception e){
```

```
9                       e.printStackTrace();
10                  }
11              }
12      );
13 }
14 private Response download(String fileName) throws IOException {
15     File f = new File(this.folderOnServer + fileName);
16     if (!f.exists()) {
17         return Response.status(Response.Status.NOT_FOUND).build();
18     } else {
19         return Response
20                 .ok(f)
21                 .header("Content-disposition", "attachment;filename=" + fileName)
22                 .header("Cache-Control", "no-cache")
23                 .build();
24     }
25 }
```

第 14 行的 download(.)函数与代码 5-21 的函数相同,无须加上任何标注。服务端异步应答就是把同步应答的功能函数放到线程池里执行,并通过 asyncResponse.resume(.)把同步应答功能函数的返回值以异步的方式答复给客户端。

客户端对应的请求函数见代码 5-40。

代码 5-40:客户端函数 download_Asyn(.)

```
1  private void download_Asyn(String remoteFileName){
2      final AsyncInvoker async = target
3              .path("file")
4              .path(remoteFileName)
5              .request(MediaType.APPLICATION_OCTET_STREAM_TYPE)
6              .async();
7      final Future<Response> responseFuture = async
8              .get(
9                      new InvocationCallback<Response>() {
10                         @Override
11                         public void completed(Response response) {
12                             if (response.getStatus() ==
                                       Response.Status.NOT_FOUND.getStatusCode()){
13                                 System.out.println("远端文件不存在:" +
                                               remoteFileName);
14                             }
15                             if (response.getStatus() ==
                                           Response.Status.OK.getStatusCode()){
16                                 String localFilePath = folderOnClient +
                                             UUID.randomUUID().toString() +
                                             "-" + remoteFileName;
```

```
17                          File localFile = new File(localFilePath);
18                          byte[] bytes = response.readEntity(byte[].class);
19                          try {
20                              OutputStream fos =
                                        new FileOutputStream(localFile);
21                              fos.write(bytes);
22                              fos.flush();
23                              fos.close();
24                          } catch (Exception e) {
25                              e.printStackTrace();
26                          }
27                          System.out.println("保存文件成功：" + localFilePath);
28                      }
29                  }
30                  @Override
31                  public void failed(Throwable throwable) {
32                      throwable.printStackTrace();
33                  }
34              }
35          );
36  }
```

在第 5 行的.request(.)之后增加调用.async()；获得一个 AsyncInvoker 对象，然后执行这个 AsyncInvoker 对象的 get(.)方法向服务端发出 GET 请求，在第 11 行的回调函数里可以得到服务端的应答，如果应答状态码为 OK，则获取字节序列并保存至文件。因为服务端功能函数的返回值为 Response 类型，所以客户端使用泛型 Future<Response>，见第 7 行；另外，第 18 行的 response.readEntity(.)只能调用一次，重复调用则发生异常，因此在回调函数的第 18 行调用了 response.readEntity(.)，则在主线程里不能再调用 response.readEntity(.)。

8. POST /file/xyz.mp4

服务端方法见代码 5-41，此方法添加于 FileResource 类。

代码 5-41：服务端函数 upload_Asyn(.)

```
1  @POST
2  @Consumes(MediaType.MULTIPART_FORM_DATA)
3  public void upload_Asyn(
                      @Suspended final AsyncResponse asyncResponse,
                      @FormDataParam("file") FormDataBodyPart bp) {
4      Server.pool.submit(
5              () ->{
6                  asyncResponse.resume(upload(bp));
```

```
7                 }
8         );
9  }
10 private Response upload(FormDataBodyPart bp) {
11     FormDataContentDisposition disposition = bp.getFormDataContentDisposition();
12     String fileName = disposition.getFileName();
13     File file = new File(this.folderOnServer +
                       UUID.randomUUID().toString() + "-" + fileName);
14     try {
15         InputStream is = bp.getValueAs(InputStream.class);
16         OutputStream fos = new FileOutputStream(file);
17         byte[] buffer = new byte[1024 * 1024];
18         int len = 0;
19         while( (len=is.read(buffer)) != -1 ){
20             fos.write(buffer, 0, len);
21             fos.flush();
22         }
23         fos.close();
24     } catch (IOException e) {
25         e.printStackTrace();
26         return Response.notModified().build();
27     }
28  return Response.ok().build();
29 }
```

第10行的upload(.)函数与代码5-23的函数相同，无须加上任何标注。服务端异步应答就是把同步应答的功能函数放到线程池里执行，并通过asyncResponse.resume(.)把同步应答功能函数的返回值以异步的方式答复给客户端。

客户端对应的请求函数见代码5-42。

代码5-42：客户端函数upload_Asyn(.)

```
1  private void upload_Asyn(String localFileName){
2      String localFilePath = folderOnClient + localFileName;
3      File localFile = new File(localFilePath);
4      if (!localFile.exists()){
5          System.out.println("本地文件不存在：" + localFilePath);
6          return;
7      }
8      FormDataMultiPart part = new FormDataMultiPart();
9      //"file"是控件命名，与服务端的@FormDataParam("file")一致
10     part.bodyPart(new FileDataBodyPart("file",localFile));
11     final AsyncInvoker async = target
12             .path("file")
13             .request(MediaType.APPLICATION_JSON_TYPE)
```

```
14              .async();
15      final Future<Response> responseFuture = async
16              .post(
17                      Entity.entity(part, MediaType.MULTIPART_FORM_DATA_TYPE),
18                      new InvocationCallback<Response>() {
19                          @Override
20                          public void completed(Response response) {
21                              if (response.getStatus() ==
                                        Response.Status.OK.getStatusCode()){
22                                  System.out.println("上传文件成功: " + localFilePath);
23                              }else{
24                                  System.out.println("上传文件失败: " + localFilePath);
25                              }
26                          }
27                          @Override
28                          public void failed(Throwable throwable) {
29                              throwable.printStackTrace();
30                          }
31                      }
32              );
33 }
```

在第13行的.request(.)之后增加调用.async()；获得一个AsyncInvoker对象，然后执行这个AsyncInvoker对象的post(.)方法向服务端发出POST请求，在第20行的回调函数里可以得到服务端的应答，如果应答状态码为OK，则说明上传文件成功。因为服务端功能函数的返回值为Response类型，所以客户端使用泛型Future<Response>，见第15行。

最后补充客户端doWork()函数，如代码5-43所示。

代码5-43：客户端函数doWork()

```
1  public void doWork(){
2       getAllStudents_Asyn();
3       getStudentById_Asyn(102);
4       getStudentById_Asyn(1020);
5       queryStudentById_Asyn(102);
6       queryStudentById_Asyn(1020);
7       addStudent_Asyn();
8       addStudent_Asyn();
9       addStudent_Asyn();
10      addStudent_Asyn();
11      deleteStudent_Asyn(102);
12      deleteStudent_Asyn(1020);
13      download_Asyn("xyz.pdf");
```

```
14        download_Asyn("xyz.mp4");
15        download_Asyn("xyz111.pdf");
16        upload_Asyn("abc.docx");
17        upload_Asyn("abc.mp4");
18        upload_Asyn("abc111.mp4");
19 }
```

运行时先启动源码中项目 JerseyServer6 的 Server 类，再启动源码中项目 JerseyClient6 的 Client 类。

5.4 案例 3：基本认证和授权

HTTP 基本认证是指通过客户端发送请求时，将用户名和口令作为身份凭证的一种登录验证方式。在发送请求时，用户名和口令合并成形如“用户名：口令”的字符串，经过 Base64 编码作为请求头部的 Authorization 属性值，如 Authorization：Basic dXNlcjp1c2VyMTIz。

本案例对应源代码 JerseyServer8 项目及 JerseyClient8 项目。

5.4.1 服务端基本框架

除了资源注册类，其他各项与 5.2.1 节相同。

代码 5-44：类 JerseyConfig

```
1  @Component
2  @ApplicationPath("/jersey")
3  public class JerseyConfig extends ResourceConfig {
4      public JerseyConfig(){
5          register(StudentResource.class);
6          register(FileResource.class);
7          register(UserAuthFilter.class);
8          register(RolesAllowedDynamicFeature.class);
9          register(MoxyXmlFeature.class);
10         register(JacksonFeature.class);
11         register(MultiPartFeature.class);
12     }
13 }
```

在 JerseyConfig 的构造器中完成资源类的注册。相比 5.2.1 节的 JerseyConfig，这里额外增加了 UserAuthFilter 和 RolesAllowedDynamicFeature 这两个类的注册。

5.4.2 客户端基本框架

客户端依赖项与5.2.2节的代码5-7相同。

客户端启动类见代码5-45。

代码5-45：类Client

```
1  public class Client {
2      public static void main(String[ ] args) {
3          Client client = new Client("user", "user123");
4          client.doWork();
5      }
6      private String account;
7      private String pwd;
8      private final String folderOnClient = "D:\\ClientFiles\\";
9      //随机选择数据格式来发送服务请求
10     private final static MediaType[ ] mediaTypes = new MediaType[ ]{
                               MediaType.APPLICATION_XML_TYPE,
                               MediaType.APPLICATION_JSON_TYPE};
11     private MediaType randomMediaType(){
12         return mediaTypes[new Random().nextInt(mediaTypes.length)];
13     }
14     private WebTarget target;
15     public Client(String account, String pwd){
16         this.account = account;
17         this.pwd = pwd;
18         ClientConfig clientConfig = new ClientConfig();
19         //使用Http头部的AUTHORIZATION字段传输用户名和口令,Base64
20         HttpAuthenticationFeature authenticationFeature =
                               HttpAuthenticationFeature.basic(account, pwd);
21         clientConfig.register(authenticationFeature);
22         clientConfig.register(MoxyXmlFeature.class);
23         clientConfig.register(JacksonFeature.class);
24         clientConfig.register(MultiPartFeature.class);
25         javax.ws.rs.client.Client rsClient = ClientBuilder.newClient(clientConfig);
26         target = rsClient.target("http://localhost:8080/jersey/");
27     }
28     public void doWork(){
39         //有待补充
30         }
31 }
```

第21行注册认证方式，第22行注册XML解码器，第23行注册JSON解码器，第24

行注册的类用于支持文件传输，第 26 行的 target(.)方法的参数指明了服务端应用路径。第 28 行的 doWork()方法有待补充。

5.4.3 服务端认证项

定义过滤器，见代码 5-46。

代码 5-46：类 UserAuthFilter

```
1  @Priority(Priorities.AUTHENTICATION)
2  public class UserAuthFilter implements ContainerRequestFilter {
3      @Inject
4      private AccountService accountService;
5      @Override
6      public void filter(ContainerRequestContext req) throws IOException {
7          String authzHeader = req.getHeaderString(HttpHeaders.AUTHORIZATION);
8          System.out.println(authzHeader);
9          final Base64.Decoder decoder = Base64.getDecoder();
10         String decoded = new String(
                   decoder.decode(authzHeader.split(" ")[1]), "UTF-8");
11         System.out.println(decoded);
12         String[] split = decoded.split(":");
13         String account = split[0];
14         String pwd = split[1];
15         //认证
16         if (account == null || !pwd.equals(accountService.getPwd(account))){
17             Response resp = Response.status(Response.Status.FORBIDDEN)
18                     .type(MediaType.APPLICATION_JSON)
19                     .entity(new String("no permission."))
20                     .build();
21             req.abortWith(resp);
22         }
23         //授权
24         SecurityContext oldContext = req.getSecurityContext();
25         req.setSecurityContext(new BasicSecurityConext(
                                   account,
                                   accountService.getRole(account),
                                   oldContext.isSecure()));
26     }
27 }
```

在过滤器中，第 16 行进行认证，若通过了认证，则在第 25 行进行授权。代码 5-47 定义了授权所用的类。

代码 5-47：类 BasicSecurityContext

```
1  public class BasicSecurityConext implements SecurityContext {
2      private String accountName;
3      private String roleName;
4      private boolean secure;
5      public BasicSecurityConext(String accountName,
                                   String roleName,
                                   boolean secure) {
6          this.accountName = accountName;
7          this.roleName = roleName;
8          this.secure = secure;
9      }
10     @Override
11     public Principal getUserPrincipal() {
12         return new Principal() {
13             @Override
14             public String getName() {
15                 return accountName;
16             }
17         };
18     }
19     @Override
20     public boolean isUserInRole(String s) {
21         return s.equals(this.roleName);
22     }
23     @Override
24     public boolean isSecure() {
25         return secure;
26     }
27     @Override
28     public String getAuthenticationScheme() {
29         return SecurityContext.BASIC_AUTH;
30     }
31 }
```

在代码 5-46 的第 4 行注入的 AccountService 类定义见代码 5-48。

代码 5-48：类 AccountService

```
1  @Component
2  public class AccountService {
3      Map<String, String> accountDB = new HashMap<>();
4      public AccountService(){
5          accountDB.put("admin", "admin123");
6          accountDB.put("user","user123");
7          accountDB.put("abc", "abc123");
```

```
8           accountDB.put("xyz","xyz123");
9       }
10      public String getPwd(String userName){
11          if (!accountDB.containsKey(userName)){
12              return "";
13          }
14          return accountDB.get(userName);
15      }
16      public String getRole(String userName){
17          String role = "COMMON_USER";
18          if (userName.equals("admin")){
19              role = "SUPER_USER";
20          }
21          if (userName.equals("xyz")){
22              role = "GUEST";
23          }
24          return role;
25      }
26  }
```

这个类模拟了从数据库取得用户名、口令及其角色的操作。这里设置了 3 种角色：SUPER_USER、COMMON_USER、GUEST，四个用户都分配了角色。

StudentResource 的访问权限只需要在各个 URI 方法上添加@RolesAllowed 标注，见代码 5-49。

代码 5-49：类 StudentResource

```
1   @Component
2   @Path("stu")
3   @PermitAll
4   public class StudentResource {
5       private static AtomicInteger idCounter = new AtomicInteger(1000);
6       private final static MediaType[] mediaTypes = new MediaType[]{
                                        MediaType.APPLICATION_XML_TYPE,
                                        MediaType.APPLICATION_JSON_TYPE};
7       private MediaType randomMediaType(){
8           return mediaTypes[new Random().nextInt(mediaTypes.length)];
9       }
10
11      @GET
12      @RolesAllowed("COMMON_USER")
13      @Produces({MediaType.APPLICATION_XML, MediaType.APPLICATION_JSON})
14      public List<Student> getAll(){
15          该函数功能同代码 5-9，此处省略
16      }
```

```
17
18      @GET
19      @Path("{id}")
20      @RolesAllowed({"SUPER_USER","GUEST"})
21      @Produces({MediaType.APPLICATION_XML, MediaType.APPLICATION_JSON})
22      public Student getOne(@PathParam("id") Integer id) {
23          该函数功能同代码 5-11,此处省略
24      }
25
26      @GET
27      @Path("query")
28      @Produces({MediaType.APPLICATION_XML, MediaType.APPLICATION_JSON})
29      public Student getOneByQuery(@QueryParam("uid") Integer id) {
30          该函数功能同代码 5-13,此处省略
31      }
32
33      @POST
34      @RolesAllowed("SUPER_USER")
35      @Consumes({MediaType.APPLICATION_XML, MediaType.APPLICATION_JSON})
36      @Produces({MediaType.APPLICATION_XML, MediaType.APPLICATION_JSON})
37      public Student createStudent(Student student){
38          该函数功能同代码 5-15,此处省略
39      }
40
41      @PUT
42      @RolesAllowed({"COMMON_USER","SUPER_USER"})
43      @Consumes({MediaType.APPLICATION_XML, MediaType.APPLICATION_JSON})
44      @Produces({MediaType.APPLICATION_XML, MediaType.APPLICATION_JSON})
45      public Response modifyStudent(Student student){
46          该函数功能同代码 5-17,此处省略
47      }
48
49      @DELETE
50      @RolesAllowed({"SUPER_USER","COMMON_USER"})
51      @Path("{id}")
52      public Response eraseStudent(@PathParam("id") Integer id){
53          该函数功能同代码 5-19,此处省略
54      }
55  }
```

FileResource 的访问权限只需要在各个 URI 方法上添加@RolesAllowed 标注，见代码 5-50。

代码 5-50：类 FileResource

```
1   @Component
2   @Path("file")
```

```
3  @PermitAll
4  public class FileResource {
5      private final String folderOnServer = "D:\\ServerFiles\\";
6
7      @GET
8      @RolesAllowed("COMMON_USER")
9      @Path("/{name}")
10     public Response download(@PathParam("name") String fileName)
       throws IOException {
11         该函数功能同代码 5-21,此处省略
12     }
13
14     @POST
15     @RolesAllowed("SUPER_USER")
16     @Consumes(MediaType.MULTIPART_FORM_DATA)
17     public Response upload(@FormDataParam("file") FormDataBodyPart bp) {
18         该函数功能同代码 5-23,此处省略
19     }
20 }
```

代码 5-49 的 StudentResource 和代码 5-50 的 FileResource 的角色权限配置实现了如表 5-2 所示的操作权限。

表 5-2 代码 5-49 的 StudentResource 和代码 5-50 的 FileResource 的角色权限配置的操作权限

URI	SUPER_USER	COMMON_USER	GUEST
GET /stu	×	√	×
GET /stu/2	√	×	√
GET /stu/query?uid=2	√	√	√
POST /stu	√	×	×
PUT /stu	√	√	×
DELETE /stu/2	√	√	×
GET /file/abc.pdf	×	√	×
POST /file/xyz.mp4	√	×	×

5.4.4 客户端认证项

客户端检测应答状态码,如果状态码为 FORBIDDEN,则表示没有操作权限,以一个函数为例来说明,见代码 5-51。

代码 5-51: 客户端函数 getAllStudents()

```
1  private List<Student> getAllStudents(){
2      MediaType mediaType = randomMediaType();
3      Response response = target
```

```
4               .path("stu")
5               .request(mediaType)
6               .get();
7       if (response.getStatus() == Response.Status.FORBIDDEN.getStatusCode()){
8           System.out.println("权限不够,无法操作!");
9           return null;
10      }
11      GenericType<List<Student>> genericType = new GenericType<>(){ };
12      List<Student> students = response.readEntity(genericType);
13      return students;
14 }
```

第7～10行用来检测是否具有操作权限。

在参照代码5-51的第7～10行修改了Client的每个请求函数(即在代码5-52的doWork()函数中要测试的那些函数)之后,最后补充客户端doWork()函数,如代码5-52所示。

代码5-52:客户端函数doWork()

```
1  public void doWork(){
2       getAllStudents();
3       getStudentById(102);
4       queryStudentById(103);
5       addStudent();
6       updateStudent(104);
7       updateStudent(1050);
8       deleteStudent(103);
9       deleteStudent(1060);
10      download("xyz.pdf");
11      download("xyz.mp4");
12      download("12345.pdf");
13      upload("abc.docx");
14      upload("abc.txt");
15      upload("12345.mp4");
16 }
```

运行时先启动源码中项目JerseyServer8的Server类,再启动源码中项目JerseyClient8的Client类。可以调整Client类的main(.)函数中用户名和口令,测试不同角色的用户。

5.5 案例4:替换某些部件

本节将逐一替换JSON解析器、Servlet容器、Tomcat服务器乃至剥离Spring。

本案例对应源代码JerseyServer9项目、JerseyServer9a项目及JerseyClient9项目。

依次替换掉 JSON 解析器、Servlet 容器、Tomcat 服务器，每一步完成后都能通过测试。这三步替换可以单独进行。

5.5.1 替换 JSON 解析器

默认情况下，Spring 用的 JSON 解析器是 Jackson，其对应的依赖如下。

```
<dependency>
    <groupId>org.glassfish.jersey.media</groupId>
    <artifactId>jersey-media-json-jackson</artifactId>
    <version>2.30.1</version>
</dependency>
```

因为在 spring-boot-starter-jersey 里已包含 jersey-media-json-jackson 依赖，故服务端不需要显式导入此依赖；Jersey 默认使用的是 Jackson 解析器，故可以不用在代码中显式注册。如果客户端要使用 Jackson 解析器，则需要导入此依赖并显式注册。

不论在服务端还是在客户端，Jackson 解析器均可以被替换掉，比如替换成 fastjson，步骤如下。

1. 服务端替换

增加 fastjson 依赖，代码如下。

```
<dependency>
    <groupId>com.alibaba</groupId>
    <artifactId>fastjson</artifactId>
    <version>1.2.73</version>
</dependency>
```

在 ResourceConfig 里注册 fastjson 解析器，代码如下。

```
register(FastJsonFeature.class);
//register(JacksonFeature.class);                    //注释掉 Jackson
```

2. 客户端替换

增加 fastjson 依赖，并删除 Jackson 依赖，代码如下。

```
<dependency>
    <groupId>com.alibaba</groupId>
    <artifactId>fastjson</artifactId>
    <version>1.2.73</version>
</dependency>
```

用 ClientConfig 注册 fastjson 解析器,并删除 jackson 的注册,代码如下。

```
clientConfig.register(FastJsonFeature.class);
//clientConfig.register(JacksonFeature.class);                    //注释掉 Jackson
```

因此,替换 JSON 解析器的操作很简单,第一步增加依赖,第二步注册 JSON 解析器。

5.5.2 替换 Servlet 容器

使用 Jersey 自带的 Servlet 容器而非 Web 服务器(Tomcat、Jetty 等)自带的 Servlet 容器。

首先,删除 ResourceConfig 上的标注,即不需要标注,代码如下。

```
1  //@Component
2  //@ApplicationPath("/jersey")
3  public class JerseyConfig extends ResourceConfig {
4      public JerseyConfig(){
5          //此处注册资源类等,(略)
6      }
7  }
```

其次,添加 ServletRegistrationBean,代码如下。

代码 5-53:类 ServletRegistrationBean

```
1  @Bean
2  public ServletRegistrationBean servletRegistrationBean() {
3      System.out.println("注册 Jersey Servlet 容器");
4      //第 1 个参数指明使用 Jersey 自带的 Servlet 容器
5      //第 2 个参数指定了 Jersey Resource 的根路径,等同于@ApplicationPath 标注
6      //第 2 个参数的 * 不能少
7      ServletRegistrationBean servletRegistrationBean =
           new ServletRegistrationBean(new ServletContainer(), "/jersey/ * ");
8      Map < String, String > map = new HashMap <>();
9      //指定 Jersey 的配置类,第 1 个参数是 Jersey 约定的,不用变
10     map.put("javax.ws.rs.Application", "cxiao.sh.cn.config.JerseyConfig");
11     //指定 Jersey Resource 包扫描路径,第 1 个参数是 Jersey 约定的,不用变
12     //可以与上一句同时使用
13     //map.put("jersey.config.server.provider.packages", "cxiao.sh.cn.another");
14     servletRegistrationBean.setInitParameters(map);
15     return servletRegistrationBean;
16 }
```

这里用到了@Bean,说明还需要使用 Spring 的自动配置功能。

5.5.3 替换 Web 服务器

用 jetty http server 替换默认的 Tomcat server。

完整的依赖如代码 5-54 所示。

代码 5-54：使用 jetty http server 时所需依赖

```
1  <dependencies>
2      <!-- Spring 自动配置功能 -->
3      <dependency>
4          <groupId>org.springframework.boot</groupId>
5          <artifactId>spring-boot-autoconfigure</artifactId>
6          <version>2.2.6.RELEASE</version>
7      </dependency>
8      <!-- 提供 jetty http Server -->
9      <!—必须排除 spring-boot-starter-jersey 默认内置的 Tomcat 才有效 -->
10     <dependency>
11         <groupId>org.springframework.boot</groupId>
12         <artifactId>spring-boot-starter-jetty</artifactId>
13         <version>2.2.6.RELEASE</version>
14     </dependency>
15     <!-- 提供 Jersey 框架,默认内置 Tomcat -->
16     <dependency>
17         <groupId>org.springframework.boot</groupId>
18         <artifactId>spring-boot-starter-jersey</artifactId>
19         <version>2.2.6.RELEASE</version>
20         <exclusions>
21             <!-- 排除 starter-jersey 内置的 tomcat  -->
22             <exclusion>
23             <groupId>org.springframework.boot</groupId>
24             <artifactId>spring-boot-starter-tomcat</artifactId>
25             </exclusion>
26         </exclusions>
27     </dependency>
28     <dependency>
29         <groupId>org.projectlombok</groupId>
30         <artifactId>lombok</artifactId>
31         <version>1.18.12</version>
32     </dependency>
33     <!-- 用 fastjson 作为 JSON 解析器,取代默认的 jackson 解析器 -->
34     <dependency>
35         <groupId>com.alibaba</groupId>
36         <artifactId>fastjson</artifactId>
```

```
37          <version>1.2.73</version>
38      </dependency>
39      <!-- 这个用于解决 LocalDate 类型与 JSON 相互转换的问题 -->
40      <dependency>
41          <groupId>com.fasterxml.jackson.datatype</groupId>
42          <artifactId>jackson-datatype-jsr310</artifactId>
43          <version>2.11.2</version>
44      </dependency>
45      <!-- 下面两个依赖用来支持 XML 数据格式 -->
46      <dependency>
47          <groupId>org.glassfish.jersey.media</groupId>
48          <artifactId>jersey-media-moxy</artifactId>
49          <version>2.30.1</version>
50      </dependency>
51      <dependency>
52          <groupId>org.glassfish.jersey.media</groupId>
53          <artifactId>jersey-media-jaxb</artifactId>
54          <version>2.30.1</version>
55      </dependency>
56      <!-- 支持文件上传和下载 -->
57      <dependency>
58          <groupId>org.glassfish.jersey.media</groupId>
59          <artifactId>jersey-media-multipart</artifactId>
60          <version>2.30.1</version>
61      </dependency>
62  </dependencies>
```

因为 SpringBoot 采用自动配置，所以导入 spring-boot-starter-jetty 依赖后无须手动配置，将自动启动 jetty 服务器并开启默认的 8080 服务端口。

spring-boot-starter-jersey 默认内置 Tomcat，若导入 spring-boot-starter-jetty 而不排除 jersey 默认内置的 Tomcat，则启动的依然是 Tomcat。

不论选用什么 HTTP 服务器，之后必然是下列二选一。

（1）用@ApplicationPath 标注 Servlet 对应的 URL 路径。

在 ResourceConfig 上作如下标注来指示 Servlet 对应的 URL 路径。

```
@Component
@ApplicationPath("/jersey")
```

（2）在 ServletRegistrationBean 中用代码自行注册 Servlet 及其 URL 路径。

```
@Bean
public ServletRegistrationBean servletRegistrationBean() {...}
```

此时应该取消在 ResourceConfig 上的标注，不过，经测试保留标注也无妨。

运行时先启动源码中项目 JerseyServer9 的 Server 类，再启动源码中项目 JerseyClient9 的 Client 类。

5.5.4 完全剥离 Spring

因为改动比较大，所以在项目 JerseyServer9a 中实现。

完整的依赖见代码 5-55。

代码 5-55：不用 Spring 时所需的全部依赖

```
1  <dependencies>
2      <!-- 下面 7 个依赖用于支持 Jetty Server 及 Servlet 容器 -->
3      <dependency>
4          <groupId>org.glassfish.jersey.core</groupId>
5          <artifactId>jersey-server</artifactId>
6          <version>2.30.1</version>
7      </dependency>
8      <dependency>
9          <groupId>org.glassfish.jersey.inject</groupId>
10         <artifactId>jersey-hk2</artifactId>
11         <version>2.30.1</version>
12     </dependency>
13     <dependency>
14         <groupId>org.glassfish.jersey.containers</groupId>
15         <artifactId>jersey-container-servlet-core</artifactId>
16         <version>2.30.1</version>
17     </dependency>
18     <dependency>
19         <groupId>org.glassfish.jersey.containers</groupId>
20         <artifactId>jersey-container-jetty-http</artifactId>
21         <version>2.30.1</version>
22     </dependency>
23     <dependency>
24         <groupId>org.eclipse.jetty</groupId>
25         <artifactId>jetty-server</artifactId>
26         <version>9.4.27.v20200227</version>
27     </dependency>
28     <dependency>
29         <groupId>org.eclipse.jetty</groupId>
30         <artifactId>jetty-servlet</artifactId>
31         <version>9.4.27.v20200227</version>
32     </dependency>
33     <dependency>
```

```
34          <groupId>org.eclipse.jetty</groupId>
35          <artifactId>jetty-util</artifactId>
36          <version>9.4.27.v20200227</version>
37      </dependency>
38      <dependency>
39          <groupId>org.projectlombok</groupId>
40          <artifactId>lombok</artifactId>
41          <version>1.18.12</version>
42      </dependency>
43      <!-- 用 fastjson 作为 JSON 解析器,不用 jackson 解析器 -->
44      <dependency>
45          <groupId>com.alibaba</groupId>
46          <artifactId>fastjson</artifactId>
47          <version>1.2.73</version>
48      </dependency>
49      <!-- 这个用于解决 LocalDate 类型与 JSON 相互转换的问题 -->
50      <dependency>
51          <groupId>com.fasterxml.jackson.datatype</groupId>
52          <artifactId>jackson-datatype-jsr310</artifactId>
53          <version>2.11.2</version>
54      </dependency>
55      <!-- 下面两个依赖用来支持 XML 数据格式 -->
56      <dependency>
57          <groupId>org.glassfish.jersey.media</groupId>
58          <artifactId>jersey-media-moxy</artifactId>
59          <version>2.30.1</version>
60      </dependency>
61      <dependency>
62          <groupId>org.glassfish.jersey.media</groupId>
63          <artifactId>jersey-media-jaxb</artifactId>
64          <version>2.30.1</version>
65      </dependency>
66      <!-- 支持文件上传和下载 -->
67      <dependency>
68          <groupId>org.glassfish.jersey.media</groupId>
69          <artifactId>jersey-media-multipart</artifactId>
70          <version>2.30.1</version>
71      </dependency>
72  </dependencies>
```

所有类(包括 ResourceConfig)上均不使用@Component、@Configure 之类的 Spring 标注。

用代码 5-56 启动 Jetty Server 并注册 Jersey Servlet。

代码 5-56：类 MyServer

```
1  public class MyServer {
2      public static void main(String[] args) {
3          MyServer myServer = new MyServer();
4          myServer.service();
5      }
6      public void service(){
7          Server server = new Server(8080);
8          ServletContextHandler context = new ServletContextHandler(
                                       ServletContextHandler.NO_SESSIONS);
9          context.setContextPath("/");
10         server.setHandler(context);
11          //配置 Servlet, 用代码配置,不用 web.xml
12         ServletHolder holder =
                   context.addServlet(ServletContainer.class, "/jersey/*");
13         holder.setInitOrder(1);
14         //.setInitParameter 可以多次调用
15         //第一个参数值是固定不变的,类似于 @ApplicationPath 标注
16         //因为 Server 是自建的,所以直接使用@ApplicationPath 无效
17         holder.setInitParameter("javax.ws.rs.Application",
                             "cxiao.sh.cn.config.JerseyConfig");
18         //第一个参数值也是固定不变的,用来指明数据包的扫描路径,暂时不用
19         //holder.setInitParameter("jersey.config.server.provider.packages",
                              "cxiao.sh.cn.another");
20         try {
21             server.start();
22             server.join();
23         } catch (Exception e) {
24             e.printStackTrace();
25         } finally {
26             server.destroy();
27         }
28     }
29 }
```

客户端不用变化。

运行时先启动源码中项目 JerseyServer9a 的 MyServer 类，再启动源码中项目 JerseyClient9 的 Client 类。

习题

编写 Jersey 程序，完成如下功能：客户端连接远程服务端，之后由使用者发送命令对服务端的文件系统进行操作，命令的顺序并不固定。

dir-列出当前文件夹内的文件及文件夹；

cd <文件夹>-改变服务端的当前文件夹；

download <远端文件名>-下载服务端的文件至本地；

upload <本地文件名>-上传本地文件至服务端；

close-结束会话。

第 6 章 SSE

SSE(Server-Sent Events)是 Web 服务器推送消息的技术规范，官网地址见 https://html.spec.whatwg.org/multipage/server-sent-events.html。Jersey 提供了一套支持 SSE 规范的 API。本章将基于 Jersey 的 SSE 模块实现订阅-发布功能和一个分布式锁。

订阅-发布功能对应源码的项目是 JerseyServer7 和 JerseyClient7，分布式锁对应源码的项目是 DistributedLockServer 和 DistributedLockClient。

6.1 SSE 概述

HTTP 协议采用“请求-应答”模式，当使用非 KeepAlive 模式时，客户端和服务端针对每个请求-应答新建一个连接，完成之后立即断开连接；当使用 KeepAlive 模式时，KeepAlive 功能使客户端到服务端的连接持续有效。如果 Jersey 服务端的服务函数上标注了@Produces (SseFeature.SERVER_SENT_EVENTS)，就表示这个服务的应答不同于普通的应答，它将是持久连接且多次的应答，这些应答就是服务端推送的事件。为了方便服务端发送这些事件以及客户端接收这些事件，Jersey 抽象出“事件通道”这一模型，在该模型中，服务端面对的是 EventOutput 及 OutboundEvent，客户端面对的是 EventInput 及 InboundEvent。因为其本质上属于 HTTP 协议的“请求-应答”模型，所以客户端无法在事件通道上作出二次请求或者对服务端推送的事件再作出“响应”，即在同一连接上无法支持复杂的交互需求。Jersey 的 SSE 流程参见图 6-1，图中用①～⑥表示操作步骤，/abc 与/xyz 可以相同，也可以不同。

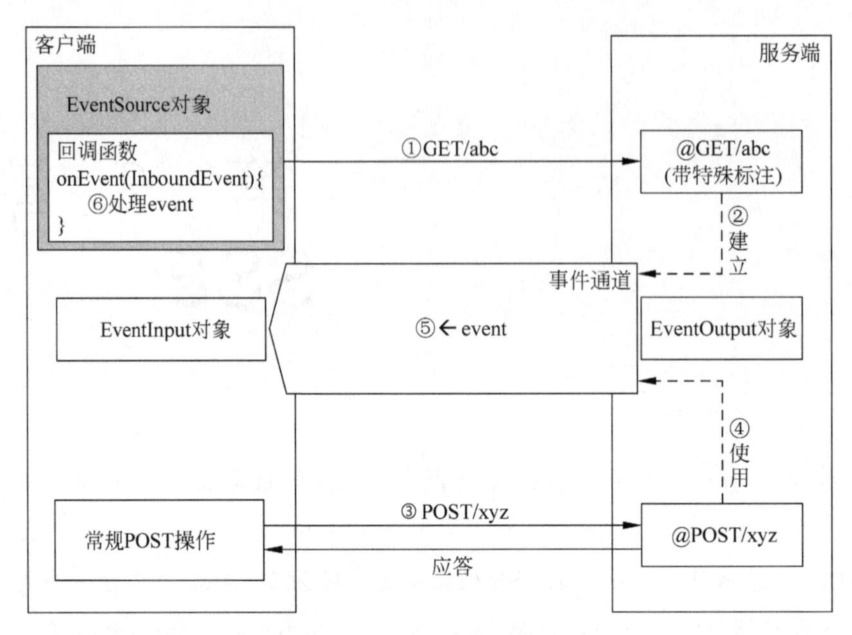

图 6-1　Jersey 的 SSE 流程图

6.2　订阅-发布功能

根据图 6-1 所示的 Jersey 的 SSE 流程，考虑这样来实现订阅-发布功能：多个客户端（即订阅者）已经通过事件通道保持与服务端的持久连接以便接收 InboundEvent，因此服务端必须管理、维护与这些事件通道对应的一组 EventOutput；任一客户端（即发布者），不论其是否已经订阅，可以通过常规的 POST 请求向服务端提交事件，服务端再通过上述所维护的一组 EventOutput 将事件分别推送至每个订阅者。而 URI/abc 的路径参数或者查询参数就是之前订阅时的订阅主题。

6.2.1　服务端代码

在本案例中，因为事件消息使用 Student 对象信息，所以依然涉及消息体编解码操作，服务端所需的依赖如代码 6-1 所示。

代码 6-1：订阅-发布时服务端依赖

```
1  <dependencies>
2      <!-- Spring 的自动配置功能 -->
3      <dependency>
4          <groupId>org.springframework.boot</groupId>
```

```
5            <artifactId> spring-boot-autoconfigure </artifactId>
6            <version> 2.2.6.RELEASE </version>
7        </dependency>
8        <!-- 提供 Jersey 框架,内置 Tomat -->
9        <dependency>
10           <groupId> org.springframework.boot </groupId>
11           <artifactId> spring-boot-starter-jersey </artifactId>
12           <version> 2.2.6.RELEASE </version>
13       </dependency>
14       <dependency>
15           <groupId> org.projectlombok </groupId>
16           <artifactId> lombok </artifactId>
17           <version> 1.18.12 </version>
18       </dependency>
19       <!-- 下面两个依赖用来支持 XML 数据格式 -->
20       <dependency>
21           <groupId> org.glassfish.jersey.media </groupId>
22           <artifactId> jersey-media-moxy </artifactId>
23           <version> 2.30.1 </version>
24       </dependency>
25       <dependency>
26           <groupId> org.glassfish.jersey.media </groupId>
27           <artifactId> jersey-media-jaxb </artifactId>
28           <version> 2.30.1 </version>
29       </dependency>
30       <!-- SSE,支持订阅-发布及广播 -->
31       <dependency>
32           <groupId> org.glassfish.jersey.media </groupId>
33           <artifactId> jersey-media-sse </artifactId>
34           <version> 2.30.1 </version>
35       </dependency>
36   </dependencies>
```

第 31～35 行导入的依赖是 Jersey 的 SSE 模块。

资源注册类见代码 6-2。

代码 6-2：类 JerseyConfig

```
1  @Component
2  @ApplicationPath("/jersey")
3  public class JerseyConfig extends ResourceConfig {
4      public JerseyConfig(){
5          register(SseSubPubResource.class);
6          register(SseBroadcastResource.class);
7          register(MoxyXmlFeature.class);
8          register(JacksonFeature.class);
```

```
9           register(SseFeature.class);
10      }
11 }
```

第5行和第6行是注册自定义的两个资源类，第9行是注册用于支持SSE的类。

资源类SseSubPubResource见代码6-3。

代码6-3：类SseSubPubResource

```
1  @Component
2  @Path("chat")
3  public class SseSubPubResource  {
4      private ConcurrentHashMap<String, List<EventOutput>> outputMap =
                                              new ConcurrentHashMap<>();
5      @GET
6      @Produces(SseFeature.SERVER_SENT_EVENTS)
7      @Path("{topic}")
8      public EventOutput register(@PathParam("topic") String topic){
9          System.out.println("主题" + topic + "被注册……");
10         EventOutput output = new EventOutput();
11         if (!outputMap.containsKey(topic)){
12             outputMap.put(topic,new ArrayList<>(Arrays.asList(output)));
13             return output;
14         }
15         outputMap.get(topic).add(output);
16         return output;
17     }
18     @POST
19     @Path("/{topic}")
20     @Consumes({MediaType.APPLICATION_XML, MediaType.APPLICATION_JSON})
21     public void postStudent(Student stu, @PathParam("topic") String topic)
       throws Exception {
22         System.out.println("主题" + topic + "被提交信息,学号" + stu.getId());
23         OutboundEvent.Builder eventBuilder = new OutboundEvent.Builder()
                                  .mediaType(MediaType.APPLICATION_JSON_TYPE);
24         OutboundEvent event = eventBuilder
25                 .id(System.nanoTime() + "")
26                 .name(topic)
27                 .data(Student.class, stu).build();
28         if (outputMap.get(topic) == null){
29             System.out.println("此主题" + topic + "未被订阅,无法发布!");
30             return;
31         }
32         List<EventOutput> outputs = outputMap.get(topic);
33         Iterator<EventOutput> iterator = outputs.iterator();
34         while (iterator.hasNext()){
```

```
35              EventOutput output = iterator.next();
36              //客户端可能关闭事件通道,所以这里要清除已关闭的通道
37              if (output.isClosed()) {
38                  iterator.remove();
39              }else{
40                  output.write(event);
41              }
42          }
43      }
44 }
```

第 4 行是服务端维护的一组 EventOutput,订阅同一主题的 EventOutput 具有相同的 Key。第 8 行是服务端提供的订阅接口,订阅时此 URI 的路径参数 topic 就是订阅主题,每个订阅者的持久通道在第 15 行被记录。第 21 行是服务端提供的发布接口,发布者发布时,此 URI 的路径参数 topic 决定了只有已经订阅了该主题的订阅者才会收到服务端的推送。发布者提交的信息(即 Student 对象 stu)在第 24～27 行被封装成 OutboundEvent 对象 event,然后在第 40 行推送至一个订阅者。通过第 34 行的循环将信息推送至该主题的所有订阅者。

Jersey 的 SSE 模块还提供了 SseBroadcaster 类用于维护一组事件通道,因此,作为资源类 SseSubPubResource 的对照,资源类 SseBroadcastResource 见代码 6-4,该类使用了 SseBroadcaster 对象来管理同一主题的多个 EventOutput 对象,其他处理与代码 6-3 并无差别。

代码 6-4:类 SseBroadcastResource

```
1  @Component
2  @Path("cast")
3  public class SseBroadcastResource {
4      //SseBroadcaster 管理一组 EventOutput
5      private ConcurrentHashMap<String, SseBroadcaster> outputMap =
                                            new ConcurrentHashMap<>();
6      @GET
7      @Produces(SseFeature.SERVER_SENT_EVENTS)
8      @Path("{topic}")
9      public EventOutput register(@PathParam("topic") String topic){
10         System.out.println("主题" + topic + "被注册……");
11         EventOutput output = new EventOutput();
12         if (!outputMap.containsKey(topic)){
13             outputMap.put(topic, new SseBroadcaster());
14         }
15         outputMap.get(topic).add(output);
16         return output;
```

```
17      }
18      @POST
19      @Path("{topic}")
20      @Consumes({MediaType.APPLICATION_XML, MediaType.APPLICATION_JSON})
21      public void postStudent(Student stu, @PathParam("topic") String topic)
        throws Exception {
22          System.out.println("主题" + topic + "被提交信息,学号" + stu.getId());
23          OutboundEvent.Builder eventBuilder = new OutboundEvent.Builder()
                                .mediaType(MediaType.APPLICATION_JSON_TYPE);
24          OutboundEvent event = eventBuilder
25                  .id(System.nanoTime() + "")
26                  .name(topic)
27                  .data(Student.class, stu).build();
28          if (outputMap.get(topic) == null){
29              System.out.println("此主题" + topic + "尚未订阅,无法发布!");
30              return;
31          }
32          outputMap.get(topic).broadcast(event);
33      }
34  }
```

服务端启动类见代码6-5。

代码6-5：类Server

```
1  @SpringBootApplication
2  public class Server {
3      public static void main(String[] args) {
4          SpringApplication.run(Server.class, args);
5      }
6  }
```

采用常规的Spring启动方式即可。

6.2.2 客户端代码

客户端所需的依赖如代码6-6所示。

代码6-6：订阅-发布时客户端依赖

```
1  <dependencies>
2      <dependency>
3          <groupId>org.projectlombok</groupId>
4          <artifactId>lombok</artifactId>
5          <version>1.18.10</version>
6      </dependency>
```

```
7        <dependency>
8            <groupId>org.glassfish.jersey.core</groupId>
9            <artifactId>jersey-client</artifactId>
10           <version>2.30.1</version>
11       </dependency>
12       <!-- 下面这三个依赖用来支持 JSON 数据格式 -->
13       <dependency>
14           <groupId>org.glassfish.jersey.inject</groupId>
15           <artifactId>jersey-hk2</artifactId>
16           <version>2.30.1</version>
17       </dependency>
18       <dependency>
19           <groupId>org.glassfish.jersey.media</groupId>
20           <artifactId>jersey-media-json-jackson</artifactId>
21           <version>2.30.1</version>
22       </dependency>
23       <dependency><!-- 这个用于解决 LocalDate 类型与 JSON 相互转换的问题 -->
24           <groupId>com.fasterxml.jackson.datatype</groupId>
25           <artifactId>jackson-datatype-jsr310</artifactId>
26           <version>2.11.2</version>
27       </dependency>
28       <!-- 下面这两个依赖及上述 jersey-hk2 三个一起用来支持 XML 数据格式 -->
29       <dependency>
30           <groupId>org.glassfish.jersey.media</groupId>
31           <artifactId>jersey-media-moxy</artifactId>
32           <version>2.30.1</version>
33       </dependency>
34       <dependency>
35           <groupId>org.glassfish.jersey.media</groupId>
36           <artifactId>jersey-media-jaxb</artifactId>
37           <version>2.30.1</version>
38       </dependency>
39       <!-- SSE,支持订阅-发布及广播 -->
40       <dependency>
41           <groupId>org.glassfish.jersey.media</groupId>
42           <artifactId>jersey-media-sse</artifactId>
43           <version>2.30.1</version>
44       </dependency>
45   </dependencies>
```

客户端类见代码 6-7。

代码 6-7：类 Client

```
1  public class Client {
2      //候选数据格式列表
```

```
3        private final static MediaType[] mediaTypes = new MediaType[]{
                                      MediaType.APPLICATION_XML_TYPE,
                                      MediaType.APPLICATION_JSON_TYPE};
4        //候选主题列表
5        private final static String[] topics = new String[]{"hello", "world", "shanghai"};
6        private static AtomicInteger idCounter = new AtomicInteger(1000);
7        private EventSource eventSource;
8        private String topicName;
9        private int cid;
10       private static MediaType randomMediaType(){
11           return mediaTypes[new Random().nextInt(mediaTypes.length)];
12       }
13       private static String randomTopicName() {
             return topics[new Random().nextInt(topics.length)];}
14       private WebTarget target;
15       public static void main(String[] args) {
16           System.out.println("测试服务端 SseSubPubResource");
17           test("http://localhost:8080/jersey/chat");
18           System.out.println("测试服务端 SseBroadcastResource");
19           test("http://localhost:8080/jersey/cast");
20           try { //防止主线程退出,因为每个客户端需要保持长连接才能接收到事件
21               Thread.currentThread().join();
22           } catch (InterruptedException e) { }
23       }
24       public static void test(String url){
25           List<Client> clients = new ArrayList<>();
26           //创建多个客户端,每个客户端都注册
27           for (int i=0;i<20;i++) {
28               clients.add(new Client(url));
29           }
30           int postTimes = 6;
31           for (int i=0; i<postTimes; i++) {
32               //随机选取一个客户端作为发布者
33               Client client = clients.get(new Random().nextInt(clients.size()));
34               client.postEvent();
35           }
36       }
37       public Client(String url){
38           //随机选择一个主题
39           topicName = randomTopicName();
40           //设置 client id,便于追踪
41           cid = idCounter.incrementAndGet();
42           ClientConfig clientConfig = new ClientConfig();
43           clientConfig.register(MoxyXmlFeature.class);
44           clientConfig.register(JacksonFeature.class);
```

```
45          clientConfig.register(SseFeature.class);
46          javax.ws.rs.client.Client rsClient = ClientBuilder.newClient(clientConfig);
47          target = rsClient.target(url);
48          eventSource = new EventSource(target.path(topicName)){
49              @Override
50              public void onEvent(InboundEvent inboundEvent) {
51                  try{
52                      Student stu = inboundEvent.readData(Student.class);
53                      System.out.println("\t\t 主题" + topicName +
                                "的订阅者 client" + cid + " 收到推送,主题:" +
                                inboundEvent.getName() + ",学号:" + stu.getId());
54                  }catch (Exception e){
55                      e.printStackTrace();
56                  }
57              }
58          };
59          System.out.println("client" + cid + " 订阅了主题 " + topicName);
60      }
61      public void postEvent(){
62          Student student = StuRepository.getStudent(new Random().nextInt(5) + 101);
63          MediaType mediaType = randomMediaType();
64          Entity<Student> studentEntity = Entity.entity(student, mediaType);
65          mediaType = randomMediaType();
66          //随意选择一个主题,并提交
67          String topic = randomTopicName();
68          System.out.println("client" + cid + " 提交,主题:" +
                        topic + ",学号:" + student.getId());
69          target.path(topic)
70                  .request(mediaType)
71                  .post(studentEntity);
72      }
73      //这里没用上
74      private void closeEvent(){
75          eventSource.close();
76      }
77  }
```

第 48 行是订阅某个主题,第 50 行的 onEvent(.)函数是收到服务端推送的消息对其进行处理。

运行时先启动源码中项目 JerseyServer7 的 Server 类,再启动源码中项目 JerseyClient7 的 Client 类。客户端控制台的显示(部分内容)如下。

测试服务端 SseSubPubResource

client1001 订阅了主题 hello

client1002 订阅了主题 world

client1003 订阅了主题 shanghai

client1004 订阅了主题 shanghai

client1005 订阅了主题 world

client1001 提交,主题：hello,学号：104

client1004 提交,主题：hello,学号：104

　　主题 hello 的订阅者 client1001 收到推送,主题：hello,学号：104

　　主题 hello 的订阅者 client1001 收到推送,主题：hello,学号：104

client1005 提交,主题：shanghai,学号：105

　　主题 shanghai 的订阅者 client1003 收到推送,主题：shanghai,学号：105

　　主题 shanghai 的订阅者 client1004 收到推送,主题：shanghai,学号：105

6.3 实现分布式锁

分布式锁具有锁的基本特点：排他性、阻塞性、可重入性。分布式锁在分布式环境下协同共享资源的使用,如图 6-2 所示为微服务架构下的应用场景。

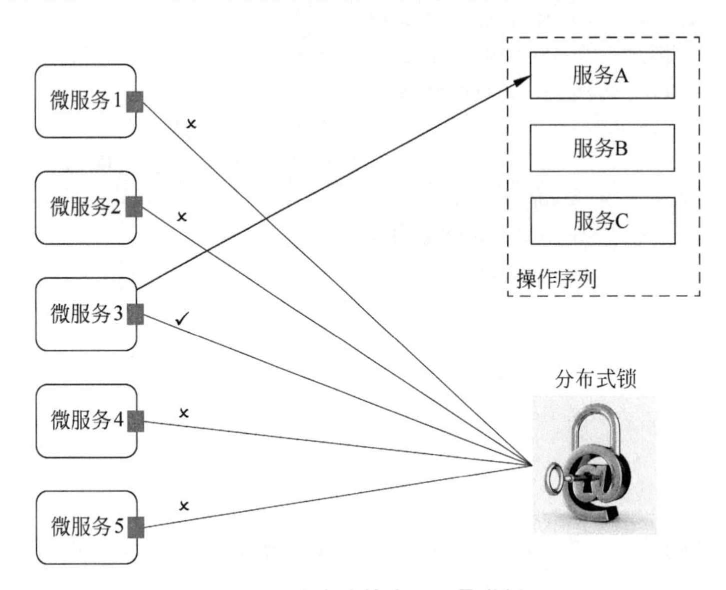

图 6-2 分布式锁应用场景举例

服务 A、B、C 构成一个操作序列,每一时刻最多只能由一个客户端执行该操作序列,现在有多个微服务都想作为客户端执行这个操作序列,这时候就需要分布式锁：这些微服务竞争这个远程锁,得到锁的那个微服务才能执行操作序列,执行完操作序列再释放锁；未得到锁的那些微服务只能阻塞等待,直到锁被释放后再次竞争锁。

分布式锁由两部分构成：一部分是锁的服务端，对应图 6-2 中“分布式锁”图案；另一部分是锁的客户端，对应图 6-2 中每个微服务上灰色小方块。分布式锁的客户端应该为使用者提供便捷的访问接口，使用者将像使用 JVM 线程锁一样来使用分布式锁，代码如下。

```
DistributedReentrantLock alock = new DistributedReentrantLock(...);
alock.lock();
此处执行操作序列
alock.unlock();
```

假设这里 DistributedReentrantLock 是分布式锁客户端为使用者提供的类，使用者无须感知锁的远程特点就能使用在远程的锁服务端。

利用 Jersey 的 SSE 特性可以实现分布式锁，锁的客户端和锁的服务端及其交互的操作流程设计如图 6-3 所示。

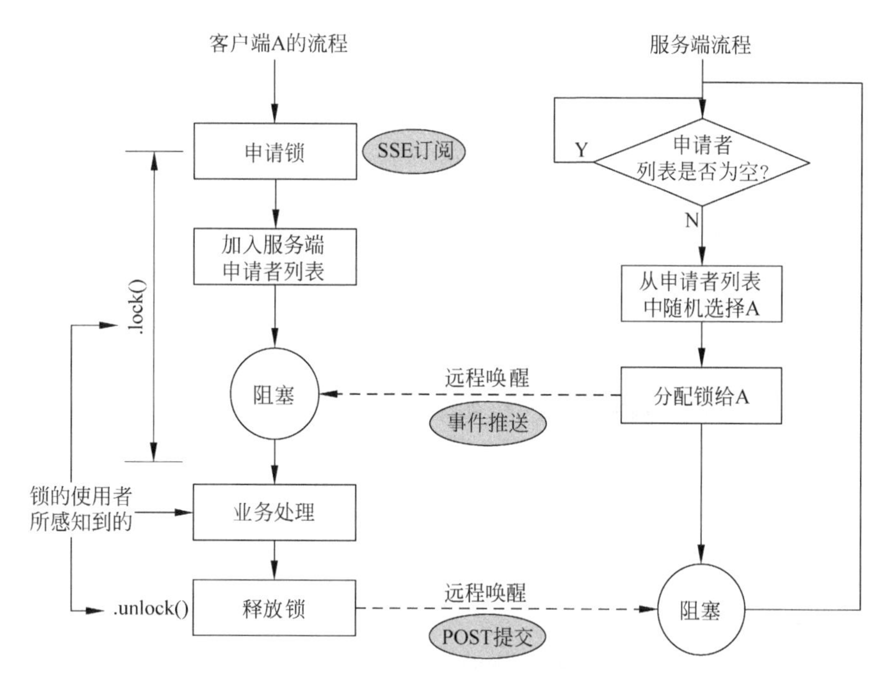

图 6-3 分布式锁客户端和服务端流程

图 6-3 中右侧的服务端每次从申请者列表中随机选择，当未选中 A 时，客户端 A 将一直阻塞；当恰巧选中 A 时，将远程唤醒客户端 A。图中灰色的椭圆形是 Jersey 的具体实现方式。

对于锁的使用者来说，在创建锁对象之后，只需调用锁对象的.lock()方法来获得锁，若.lock()方法返回则表示成功获得锁，否则该方法将一直阻塞；之后执行业务处理；最后调用锁对象的.unlock()方法释放锁。

6.3.1 分布式锁服务端

服务端所需的依赖见代码6-8。

代码6-8：分布式锁服务端依赖

```
1  <dependencies>
2      <!-- Spring 的自动配置功能 -->
3      <dependency>
4          <groupId>org.springframework.boot</groupId>
5          <artifactId>spring-boot-autoconfigure</artifactId>
6          <version>2.2.6.RELEASE</version>
7      </dependency>
8      <!-- 提供 Jersey 框架,内置 Tomat -->
9      <dependency>
10         <groupId>org.springframework.boot</groupId>
11         <artifactId>spring-boot-starter-jersey</artifactId>
12         <version>2.2.6.RELEASE</version>
13     </dependency>
14     <dependency>
15         <groupId>org.projectlombok</groupId>
16         <artifactId>lombok</artifactId>
17         <version>1.18.12</version>
18     </dependency>
19     <!-- 下面两个依赖用来支持 XML 数据格式 -->
20     <dependency>
21         <groupId>org.glassfish.jersey.media</groupId>
22         <artifactId>jersey-media-moxy</artifactId>
23         <version>2.30.1</version>
24     </dependency>
25     <dependency>
26         <groupId>org.glassfish.jersey.media</groupId>
27         <artifactId>jersey-media-jaxb</artifactId>
28         <version>2.30.1</version>
29     </dependency>
30     <!-- SSE,支持订阅-发布及广播 -->
31     <dependency>
32         <groupId>org.glassfish.jersey.media</groupId>
33         <artifactId>jersey-media-sse</artifactId>
34         <version>2.30.1</version>
35     </dependency>
36 </dependencies>
```

第31～35行导入的依赖是Jersey的SSE模块。

资源注册类见代码 6-9。

代码 6-9：类 JerseyConfig

```
1  @Component
2  @ApplicationPath("/jersey")
3  public class JerseyConfig extends ResourceConfig {
4      public JerseyConfig(){
5          register(LockResource.class);
6          register(MoxyXmlFeature.class);
7          register(JacksonFeature.class);
8          register(SseFeature.class);
9      }
10 }
```

第 5 行是注册自定义的 LockResource 类，这个类是分布式锁服务端的核心；第 8 行是注册用于支持 SSE 的类。

资源类 LockResource 见代码 6-10。

代码 6-10：类 LockResource

```
1  @Component
2  @Path("lock")
3  public class LockResource implements Runnable{
4      private ExecutorService fixPool = Executors.newCachedThreadPool();
5      private ConcurrentHashMap<String, EventOutput> outputMap =
                                                     new ConcurrentHashMap<>();
6      private final static MediaType[] mediaTypes = new MediaType[]{//只用 JSON
                                     MediaType.APPLICATION_JSON_TYPE};
7      private static MediaType randomMediaType(){
8          return mediaTypes[new Random().nextInt(mediaTypes.length)];
9      }
10     CyclicBarrier cb = new CyclicBarrier(2);
11     public LockResource(){
12         fixPool.submit(this);
13         fixPool.shutdown();
14     }
15     @GET
16     @Produces(SseFeature.SERVER_SENT_EVENTS)
17     @Path("{uuid}")
18     public EventOutput register(@PathParam("uuid") String uuid){
19         System.out.println("client " + uuid + " 试图申请锁……");
20         if (!outputMap.containsKey(uuid)){
21             EventOutput output = new EventOutput();
22             outputMap.put(uuid, output);
23         }
```

```
24            return outputMap.get(uuid);
25        }
26        @POST
27        @Path("{uuid}")
28        @Consumes({MediaType.APPLICATION_XML, MediaType.APPLICATION_JSON})
29        public void postStudent(String uid, @PathParam("uuid") String uuid)
            throws Exception {
30            System.out.println("client " + uuid + " 释放锁!");
31            //从 Map 中删除此 uuid 项,并关闭
32            if (outputMap.containsKey(uuid)){
33                EventOutput output = outputMap.remove(uuid);
34                output.close();
35            }
36            cb.await();
37        }
38        @Override
39        public void run() { //单一线程
40            while(true){
41                //从现有的申请者中随机地挑选一个,为其发放锁
42                List<String> keyList = new ArrayList<String>(outputMap.keySet());
43                if (keyList.size()>0) {
44                    String randKey = keyList.get(new Random().nextInt(keyList.size()));
45                    EventOutput output = outputMap.get(randKey);
46                    OutboundEvent.Builder eventBuilder =
                                             new OutboundEvent.Builder()
                                             .mediaType(randomMediaType());
47                    OutboundEvent event = eventBuilder.id(randKey)
48                            .name(new Date().toString())
49                            .data(String.class, randKey).build();
50                    try {
51                        //精准推送给客户端,让其获得锁。避免惊群效应
52                        output.write(event);
53                        System.out.println("client " + randKey + " 获得了锁!");
54                        //设置 await(.)的超时参数可以消除客户端死机造成的死锁
55                        cb.await();  //发放了一个锁就停
56                    } catch (Exception ex) {
57                        ex.printStackTrace();
58                        //如果设置了 cb.await(.)的超时参数,则需要增加以下处理
59                        if (ex instanceof TimeoutException){
60                            if (outputMap.containsKey(randKey)){
61                                EventOutput op = outputMap.remove(randKey);
62                                try {
63                                    op.close();
64                                }catch (IOException ioe){ }
65                            }
```

```
66                    }
67                }
68            }
69        }
70    }
71 }
```

锁服务端启动类见代码 6-11。

代码 6-11：类 Server

```
1  @SpringBootApplication
2  public class Server {
3      public static void main(String[] args) {
4          SpringApplication.run(Server.class, args);
5      }
6  }
```

采用常规的 Spring 启动方式即可。

6.3.2 分布式锁客户端

客户端所需的依赖如代码 6-12。

代码 6-12：分布式锁客户端依赖

```
1  <dependencies>
2      <dependency>
3          <groupId>org.projectlombok</groupId>
4          <artifactId>lombok</artifactId>
5          <version>1.18.10</version>
6      </dependency>
7      <dependency>
8          <groupId>org.glassfish.jersey.core</groupId>
9          <artifactId>jersey-client</artifactId>
10         <version>2.30.1</version>
11     </dependency>
12     <dependency>
13         <groupId>org.glassfish.jersey.inject</groupId>
14         <artifactId>jersey-hk2</artifactId>
15         <version>2.30.1</version>
16     </dependency>
17     <dependency>
18         <groupId>org.glassfish.jersey.media</groupId>
19         <artifactId>jersey-media-json-jackson</artifactId>
20         <version>2.30.1</version>
```

```
21        </dependency>
22        <dependency><!-- 这个用于解决LocalDate类型与JSON相互转换的问题 -->
23            <groupId>com.fasterxml.jackson.datatype</groupId>
24            <artifactId>jackson-datatype-jsr310</artifactId>
25            <version>2.11.2</version>
26        </dependency>
27        <dependency>
28            <groupId>org.glassfish.jersey.media</groupId>
29            <artifactId>jersey-media-moxy</artifactId>
30            <version>2.30.1</version>
31        </dependency>
32        <dependency>
33            <groupId>org.glassfish.jersey.media</groupId>
34            <artifactId>jersey-media-jaxb</artifactId>
35            <version>2.30.1</version>
36        </dependency>
37        <!-- SSE,支持订阅-发布及广播 -->
38        <dependency>
39            <groupId>org.glassfish.jersey.media</groupId>
40            <artifactId>jersey-media-sse</artifactId>
41            <version>2.30.1</version>
42        </dependency>
43    </dependencies>
```

分布式锁客户端见代码6-13。

代码6-13：类DistributedLock

```
1   public class DistributedLock implements Lock {
2       private final static MediaType[] mediaTypes = new MediaType[]{ //只用JSON
                                       MediaType.APPLICATION_JSON_TYPE};
3       private static MediaType randomMediaType(){
4           return mediaTypes[new Random().nextInt(mediaTypes.length)];
5       }
6       private String sURL;
7       private EventSource eventSource;
8       private String uuid = null;
9       private WebTarget target;
10      public DistributedLock(String sURL){
11          this.sURL = sURL;
12          if (this.sURL == null) {
13              this.sURL = "http://localhost:8080/jersey/lock";
14          }
15          ClientConfig clientConfig = new ClientConfig();
16          clientConfig.register(MoxyXmlFeature.class);
17          clientConfig.register(JacksonFeature.class);
```

```
18          clientConfig.register(SseFeature.class);
19          javax.ws.rs.client.Client rsClient = ClientBuilder.newClient(clientConfig);
20          target = rsClient.target(this.sURL);
21      }
22      @Override
23      public void lock() { //阻塞直至得到锁为止
24          //每一次 lock()都是重新订阅
25          CountDownLatch latch = new CountDownLatch(1);
26          if (uuid == null){
27              uuid = UUID.randomUUID().toString().replaceAll("-", "");
28          }
29          eventSource = new EventSource(target.path(uuid)){
30              @Override
31              public void onEvent(InboundEvent inboundEvent) {
32                  try{
33                      //返回的是 uuid
34                      String msg = inboundEvent.readData(String.class);
35                      if (msg.equals(uuid)){
36                          latch.countDown();
37                          System.out.println("client " + uuid + " 获得锁!");
38                      }
39                  }catch (Exception e){
40                      e.printStackTrace();
41                  }
42              }
43          };
44          System.out.println("client " + uuid + " 申请锁……");
45          try {
46              latch.await();
47          }catch (Exception ex){
48              ex.printStackTrace();
49          }
50      }
51      @Override
52      public void lockInterruptibly() throws InterruptedException {
53      }
54      @Override
55      public boolean tryLock() {
56          return false;
57      }
58      @Override
59      public boolean tryLock(long time, TimeUnit unit) throws InterruptedException {
60          return false;
61      }
62      @Override
```

```
63      public void unlock() {
64          if (uuid == null){
65              return;
66          }
67          MediaType mediaType = randomMediaType();
68          Entity<String> msgEntity = Entity.entity(uuid, mediaType);
69          mediaType = randomMediaType();
70          target.path(uuid)
71                  .request(mediaType)
72                  .post(msgEntity);
73          System.out.println("client " + uuid + " 释放了锁!");
74          eventSource.close();
75      }
76      @Override
77      public Condition newCondition() {
78          return null;
79      }
80  }
```

这里只实现了接口 java.util.concurrent.locks.Lock 的 lock()方法和 unlock()方法，其他方法采用默认实现。

代码 6-13 的类 DistributedLock 没有考虑锁的可重入性，因此再定义一个可重入的分布式锁 DistributedReentrantLock，见代码 6-14。该类继承了 DistributedLock。

代码 6-14：类 DistributedReentrantLock

```
1   public class DistributedReentrantLock extends DistributedLock {
2       private ThreadLocal<Integer> reentrantCount = new ThreadLocal<>();
3       public DistributedReentrantLock(String sURL){
4           super(sURL);
5           reentrantCount.set(null);
6       }
7       @Override
8       public void lock() {
9           //可重入处理
10          if (this.reentrantCount.get()!= null){
11              int count = this.reentrantCount.get();
12              if (count > 0){
13                  this.reentrantCount.set(++count);
14                  return;
15              }
16          }
17          super.lock();
18          reentrantCount.set(1);
```

```
19      }
20      @Override
21      public void unlock() {
22          //重入时的退出处理
23          if (this.reentrantCount.get()!= null){
24              int count = this.reentrantCount.get();
25              if (count > 1){
26                  this.reentrantCount.set( -- count);
27                  return;
28              }else{
29                  this.reentrantCount.set(null);
30              }
31          }
32          super.unlock();
33      }
34 }
```

6.3.3 分布式锁的使用

类 Client 作为锁的使用者，对锁进行测试，参见代码 6-15。

代码 6-15：类 Client

```
1  public class Client implements Runnable{
2      //此线程池用来专门运行当前 Client 的 run()方法
       //所有 Client 共享静态变量 number
3      private final ExecutorService fixPool = Executors.newCachedThreadPool();
4      private static int number = 0;
5      private final static MediaType[] mediaTypes = new MediaType[]{
                                          MediaType.APPLICATION_JSON_TYPE};
6      private static MediaType randomMediaType(){
7          return mediaTypes[new Random().nextInt(mediaTypes.length)];
8      }
9      private CountDownLatch latch;
10     public static void main(String[] args) {
11         int concurrency = 200;
12         CountDownLatch myLatch = new CountDownLatch(concurrency);
13         for (int i = 0;i < concurrency;i++) {
14             new Client(myLatch);
15             //模拟并发的场景
16             myLatch.countDown();
17             System.out.println((i + 1) + " client 准备好了……");
18         }
19     }
```

```
20      public Client(CountDownLatch latch) {
21          this.latch = latch;
22          fixPool.submit(this);
23          //在执行 shutdown()后,已提交的任务会继续处理而不允许再提交新的任务
24          //所以 fixPool 不能是静态变量
25          fixPool.shutdown();
26      }
27      @Override
28      public void run(){
29          try {
30              latch.await();
31          }catch (Exception ex) {
32              ex.printStackTrace();
33          }
34          //两个类,一个是不可重入锁,另一个是可重入锁
35          //测试不可重入锁时要把嵌套的 lock()删除
36          //初始化参数指明提供分布式锁服务的地址
37          DistributedReentrantLock alock =
                    new DistributedReentrantLock("http://localhost:8080/jersey/lock");
38          alock.lock();
39          number = number + 1;
40          alock.lock(); //测试锁的可重入性
41          System.out.println("目前的数值是:");
42          System.out.println(number);
43          alock.unlock();
44          alock.unlock();
45      }
46  }
```

运行时先启动源码中项目 DistributedLockServer 的 Server 类,再启动源码中项目 DistributedLockClient 的 Client 类。

习题

以 6.2 节的案例为基础,编写聊天室软件。

第 7 章 实现RPC框架

RPC(Remote Procedure Call)称为远程过程调用。RPC 采用客户端-服务端结构,通过请求-应答这种消息模式来实现。近些年来,因为分布式系统、服务化以及微服务的兴起,RPC 因具有服务化、可重用、系统级互操作等优点再度成为一个热点。

本章将设计和实现一个 RPC 框架,对应源代码 RpcServer 项目及 RpcClient 项目。

7.1 RPC 框架概述

一般来说,RPC 的具体操作流程包括如下步骤。

(1) 客户端调用 RPC 本地代理,就像调用本地方法一样,传递参数。

(2) 本地代理把调用的方法名、参数值等信息编组为消息,向服务端发送。

(3) 服务端接收到消息后将其解组为方法名、参数值等。

(4) 服务端调用自身的服务实例的方法,传入参数,得到结果。

(5) 服务端以反方向的相同步骤将结果响应给客户端。

RPC 框架对参数编组、消息解组、底层网络通信等步骤进行了封装,带来的便捷是开发者可以直接在其基础上进行只需专注于服务功能的代码编写。目前,远程过程多以(微)服务的形式出现,服务发布者向注册中心注册所发布的服务,注册中心相当于服务信息数据库,服务调用者从注册中心发现所需的服务信息,然后根据此信息向服务发布者进行远程过程调用。图 7-1 显示了使用 RPC 框架进行服务消费时的流程和模块。

图 7-1 中 Request 对象、Response 对象的类型定义分别见代码 7-1 和代码 7-2。

代码 7-1:类 Request

```
1  @Data
2  public class Request implements Serializable {
3      private String serviceName;
4      private String version;
5      private String method;
```

```
6       private Map<String, String> headers = new HashMap<String, String>();
7       private Class<?>[] prameterTypes;
8       private Object[] parameters;
9       public String getHeader(String name) {
10          return this.headers == null ? null : this.headers.get(name);
11      }
12      public void setHeader(String key, String value){
13          this.headers.put(key, value);
14      }
15 }
```

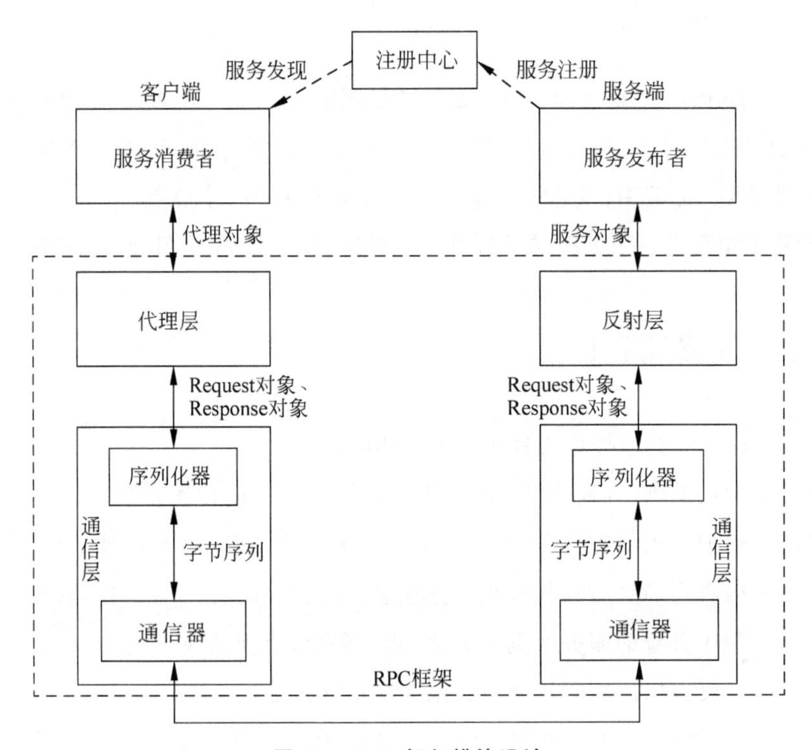

图 7-1　RPC 框架模块设计

类 Request 对调用的服务名、方法及参数值等信息进行了封装，这个对象由客户端生成，并发送至服务端。

代码 7-2：类 Response

```
1  @Data
2  public class Response implements Serializable {
3      private Map<String, String> headers = new HashMap<>();
4      private Status status;
5      private Object return Value;
6      private Exception exception;
7      public Response() {   }
8      public Response(Status status) {
```

```
9           this.status = status;
10      }
11      public String getHeader(String name) {
12          return this.headers == null ? null : this.headers.get(name);
13      }
14      public void setHeader(String key, String value) {
15          this.headers.put(key, value);
16      }
17 }
```

类 Response 对服务端执行方法调用的返回值等进行了封装,这个对象由服务端生成,并发送至客户端。

7.2 框架的客户端设计

框架客户端导入的依赖如代码 7-3 所示。

代码 7-3:框架客户端的依赖项

```
1 <dependencies>
2     <dependency>
3         <groupId>com.thoughtworks.xstream</groupId>
4         <artifactId>xstream</artifactId>
5         <version>1.4.12</version>
6     </dependency>
7     <dependency>
8         <groupId>io.netty</groupId>
9         <artifactId>netty-all</artifactId>
10        <version>4.1.51.Final</version>
11    </dependency>
12    <dependency>
13        <groupId>org.projectlombok</groupId>
14        <artifactId>lombok</artifactId>
15        <version>1.18.12</version>
16    </dependency>
17 </dependencies>
```

第 2～6 行是用于支持对象的 XML 格式化。

7.2.1 序列化器

框架的客户端发送 Request 对象并接收 Response 对象,需要对 Request 对象作序列化而对 Response 对象作反序列化,因此设计如代码 7-4 所示的接口。

代码 7-4：接口 IClientSerializer

```
1  public interface IClientSerializer<M, U> {
2      //把第一个泛型 M 序列化,把第二个泛型 U 反序列化
3      byte[] marshalling(M t) throws Exception;
4      U unmarshalling(byte[] data) throws Exception;
5  }
```

类 ClientJavaSerializer 是 IClientSerializer 的一种实现，采用了 Java 自带的序列化/反序列化方法，见代码 7-5。

代码 7-5：类 ClientJavaSerializer

```
1  public class ClientJavaSerializer implements IClientSerializer<Request, Response>{
2      @Override
3      public byte[] marshalling(Request request) throws Exception {
4          return objSerializableToByteArray(request);
5      }
6      @Override
7      public Response unmarshalling(byte[] data) throws Exception {
8          return (Response) byteArrayToObjSerializable(data);
9      }
10     //反序列化,将字节转换成对象,Object 必须实现 Serializable 接口
11     private static Object byteArrayToObjSerializable(byte[] bytes) throws Exception{
12         ByteArrayInputStream byteArrayInputStream =
                                  new ByteArrayInputStream(bytes);
13         ObjectInputStream objectInputStream =
                                  new ObjectInputStream(byteArrayInputStream);
14         Object obj = objectInputStream.readObject();
15         objectInputStream.close();
16         byteArrayInputStream.close();
17         return obj;
18     }
19     //序列化,将对象转换成字节,Object 必须实现 Serializable 接口
20     private static byte[] objSerializableToByteArray(Object objSerializable)
         throws Exception{
21         ByteArrayOutputStream byteArrayOutputStream =
                                  new ByteArrayOutputStream();
22         ObjectOutputStream objectOutputStream =
                                  new ObjectOutputStream(byteArrayOutputStream);
23         objectOutputStream.writeObject(objSerializable);
24         byte[] bytes = byteArrayOutputStream.toByteArray();
25         objectOutputStream.close();
26         byteArrayOutputStream.close();
27         return bytes;
28     }
29 }
```

类 ClientXMLSerializer 是 IClientSerializer 的另一种实现，采用 XML 格式序列化方法，见代码 7-6。

代码 7-6：类 ClientXMLSerializer

```
1  public class ClientXMLSerializer implements IClientSerializer<Request, Response> {
2      private XStream xStream;
3      public ClientXMLSerializer(){
4          xStream = new XStream();
5          xStream.alias(Request.class.getSimpleName(), Request.class);
6          xStream.alias(Response.class.getSimpleName(), Response.class);
7      }
8      @Override
9      public byte[] marshalling(Request request) throws Exception {
10         String xml = xStream.toXML(request);
11         return xml.getBytes("UTF-8");
12     }
13     @Override
14     public Response unmarshalling(byte[] data) throws Exception {
15         String xml = new String(data, "UTF-8");
16         return (Response) xStream.fromXML(xml);
17     }
18 }
```

7.2.2 代理层

在代理层设计类 ClientStubProxyFactory，见代码 7-7。

代码 7-7：类 ClientStubProxyFactory

```
1  @Data
2  public class ClientStubProxyFactory {
3      private ServiceInfoDiscoverer sid;
4      private NetClient<Request,Response> netClient;
5      private Map<String, IClientSerializer<Request,Response>> supportingSerializers;
6      //key 的形式为 interfName:version
7      private Map<String, Object> proxyCache = new HashMap<>();
8      public <T> T getProxy(Class<T> interf, String version) {
9          String key = interf.getName() + ":" + version;
10         T obj = (T) this.proxyCache.get(key);
11         if (obj == null) {
12             obj = (T) Proxy.newProxyInstance(interf.getClassLoader(),
13                             new Class<?>[] { interf },
14                             new ClientStubInvocationHandler(interf, version));
```

```
15              this.proxyCache.put(key, obj);
16          }
17          return obj;
18      }
19      private class ClientStubInvocationHandler implements InvocationHandler {
20          private Class<?> interf;
21          private String version;
22          public ClientStubInvocationHandler(Class<?> interf, String version) {
23              super();
24              this.interf = interf;
25              this.version = version;
26          }
27          @Override
28          public Object invoke(Object proxy, Method method, Object[] args)
              throws Throwable {
29              if (method.getName().equals("toString")) {
30                  return proxy.getClass().toString();
31              }
32              if (method.getName().equals("hashCode")) {
33                  return 0;
34              }
35              //1. 获得服务信息
36              String serviceName = this.interf.getName();
37              List<ServiceInfo> sinfos = sid.getServiceInfo(serviceName, this.version);
38              if (sinfos == null || sinfos.size() == 0) {
39                  throw new Exception("远程服务不存在!");
40              }
41              //2. 随机选择一个服务提供者(软负载均衡)
42              ServiceInfo sinfo = sinfos.get(new Random().nextInt(sinfos.size()));
43              //3. 构造 Request 对象
44              Request req = new Request();
45              req.setServiceName(sinfo.getInterfName());
46              req.setMethod(method.getName());
47              req.setPrameterTypes(method.getParameterTypes());
48              req.setParameters(args);
49              req.setVersion(sinfo.getVersion());
50              //4. 获得对应的序列化器
51              IClientSerializer<Request, Response> serializing =
                                  supportingSerializers.get(sinfo.getSerializingType());
52              //5. 发送请求并得到应答
53              Response rsp =
                          netClient.sendRequest(req, sinfo.getAddress(), serializing);
54              if (rsp.getException() != null) {
55                  throw rsp.getException();
56              }
```

```
57                return rsp.getReturnValue();
58            }
59        }
60 }
```

第8～18行是根据接口(即发布的服务)获得该接口的代理对象。当调用该代理对象的方法时,真正的操作委托给第14行创建的ClientStubInvocationHandler对象来完成。ClientStubInvocationHandler类的定义从第19行开始,第28行的invoke(.)函数完成当代理对象的方法被调用时真正的操作。对于要实现的RPC框架而言,invoke(.)要做的事包括根据要消费的服务来构造Request对象,之后将此对象发送至服务发布方以获得远程调用的结果Response对象,Request对象的序列化操作以及Response对象的反序列化操作是通过第51行选取的序列化器来完成的,最后第57行从Response对象里得到远程函数的返回值。

7.2.3 通信层

通信层设计为接口NetClient,见代码7-8。序列化器设计为通信层的一部分,因此通信层把Request对象作为发送对象,并得到返回的Response对象,这里分别用泛型T和U代表Request和Response。

代码7-8:接口NetClient

```
1  public interface NetClient<T, U> {
2      U sendRequest(T request, String sAddress, IClientSerializer<T, U> serializing)
       throws Throwable;
3  }
```

类NettyNetClient是接口NetClient的一种实现,见代码7-9。该类用Netty实现通信器。

代码7-9:类NettyNetClient

```
1  public class NettyNetClient implements NetClient<Request, Response> {
2      @Override
3      public Response sendRequest(Request request,
                                   String sAddress,
                                   IClientSerializer<Request, Response> serializing)
       throws Throwable {
4          String[] addInfoArray = sAddress.split(":");  //sAddress为ip:port格式
5          SendHandler<Request, Response> sendHandler =
                                   new SendHandler<>(request, serializing);
6          Response response = null;
7          //配置Netty客户端
8          EventLoopGroup group = new NioEventLoopGroup();
9          try {
10             Bootstrap b = new Bootstrap();
```

```
11                b.group(group)
12                        .channel(NioSocketChannel.class)
13                        .remoteAddress(new InetSocketAddress(
                                                    addInfoArray[0],
                                                    Integer.valueOf(addInfoArray[1])))
14                        .option(ChannelOption.TCP_NODELAY, true)
15                        .handler(new ChannelInitializer<SocketChannel>() {
16                            @Override
17                            public void initChannel(SocketChannel ch)
                                throws Exception {
18                                ch.pipeline().addLast(
                                            new LengthFieldBasedFrameDecoder(
                                                    Integer.MAX_VALUE, 0, 4, 0, 4));
19                                ch.pipeline().addLast(sendHandler);
20                                ch.pipeline().addLast(new AttachHeaderHandler());
21                            }
22                        });
23                //启动 Netty 客户端连接
24                b.connect().sync();
25                response = sendHandler.getResponse();
26            } finally {
27                //释放线程组资源
28                group.shutdownGracefully();
29            }
30            return response;
31        }
32        //T 是最初发送出去的类型,U 是最后接收到的类型
33        private class SendHandler<T, U> extends ChannelInboundHandlerAdapter {
34            private T request;
35            private U response = null;
36            private IClientSerializer<T, U> serializing;
37            private CountDownLatch cdl;
38            public SendHandler(T request, IClientSerializer<T, U> serializing) {
39                cdl = new CountDownLatch(1);
40                this.request = request;
41                this.serializing = serializing;
42            }
43            @Override
44            public void channelActive(ChannelHandlerContext ctx) throws Exception {
45                byte[] data = this.serializing.marshalling(this.request);
46                ByteBuf reqBuf = Unpooled.buffer(data.length);
47                reqBuf.writeBytes(data);
48                ctx.channel().write(reqBuf);
49            }
50            public U getResponse() {
```

```
51              try {
52                  cdl.await();
53              } catch (InterruptedException e) {
54                  e.printStackTrace();
55              }
56              return this.response;
57          }
58          @Override
59          public void channelRead(ChannelHandlerContext ctx, Object msg)
            throws Exception {
60              ByteBuf msgBuf = (ByteBuf) msg;
61              byte[] bytes = new byte[msgBuf.readableBytes()];
62              msgBuf.readBytes(bytes);
63              this.response = this.serializing.unmarshalling(bytes);
64              cdl.countDown();
65          }
66          @Override
67          public void exceptionCaught(ChannelHandlerContext ctx, Throwable cause) {
68              cause.printStackTrace();
69              ctx.close();
70          }
71      }
72 }
```

第25行是在Netty连接建立之后马上去取得服务端的应答对象response，但是服务端方法的执行可能比较耗时，不能马上给出应答，因此getResponse()方法必须等待，直到收到服务端的应答再返回值。第50～57行是该方法的定义，方法里用了一个CountDownLatch，第52行的cdl.await()只有在第64行的cdl.countDown()执行之后才会结束从而获得有效的response对象，而第64行的cdl.countDown()发生在客户端接收到服务端的应答对象response之后，见第60～63行。

客户端通信层的类图见图7-2。

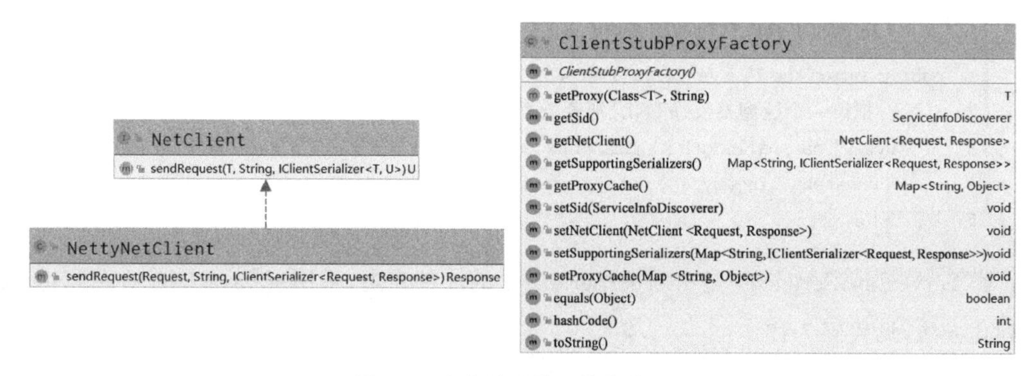

图7-2 客户端通信层的类关系图

7.3 框架的服务端设计

框架服务端导入的依赖如代码 7-10 所示。

代码 7-10：框架服务端的依赖项

```
1 <dependencies>
2     <dependency>
3         <groupId>com.thoughtworks.xstream</groupId>
4         <artifactId>xstream</artifactId>
5         <version>1.4.12</version>
6     </dependency>
7     <dependency>
8         <groupId>io.netty</groupId>
9         <artifactId>netty-all</artifactId>
10        <version>4.1.51.Final</version>
11    </dependency>
12    <dependency>
13        <groupId>org.projectlombok</groupId>
14        <artifactId>lombok</artifactId>
15        <version>1.18.12</version>
16    </dependency>
17 </dependencies>
```

第 2～6 行用于支持对象的 XML 格式化。

7.3.1 序列化器

框架的服务端接收 Request 对象并发送 Response 对象，需要对 Request 对象作反序列化，而对 Response 对象作序列化，因此设计如代码 7-11 所示的接口。

代码 7-11：接口 IServerSerializer

```
1 public interface IServerSerializer<U, M> {
2     //把第一个泛型 U 反序列化，把第二个泛型 M 序列化
3     byte[] marshalling(M m) throws Exception;
4     U unmarshalling(byte[] data) throws Exception;
5 }
```

类 ServerJavaSerializer 是 IServerSerializer 的一种实现，采用了 Java 自带的序列化/反序列化方法，见代码 7-12。

代码 7-12：类 ServerJavaSerializer

```
1  public class ServerJavaSerializer implements IServerSerializer<Request, Response>{
2      @Override
3      public byte[] marshalling(Response response) throws Exception {
4          return objSerializableToByteArray(response);
5      }
6      @Override
7      public Request unmarshalling(byte[] data) throws Exception {
8          return (Request)byteArrayToObjSerializable(data);
9      }
10     //反序列化,将字节转换成对象,Object 必须实现 Serializable 接口
11     private static Object byteArrayToObjSerializable(byte[] bytes) throws Exception{
12         ByteArrayInputStream byteArrayInputStream =
                               new ByteArrayInputStream(bytes);
13         ObjectInputStream objectInputStream =
                               new ObjectInputStream(byteArrayInputStream);
14         Object obj = objectInputStream.readObject();
15         objectInputStream.close();
16         byteArrayInputStream.close();
17         return obj;
18     }
19     //序列化,将对象转换成字节,Object 必须实现 Serializable 接口
20     private static byte[] objSerializableToByteArray(Object objSerializable)
       throws Exception{
21         ByteArrayOutputStream byteArrayOutputStream =
                               new ByteArrayOutputStream();
22         ObjectOutputStream objectOutputStream =
                               new ObjectOutputStream(byteArrayOutputStream);
23         objectOutputStream.writeObject(objSerializable);
24         byte[] bytes = byteArrayOutputStream.toByteArray();
25         objectOutputStream.close();
26         byteArrayOutputStream.close();
27         return bytes;
28     }
29 }
```

类 ServerXMLSerializer 是 IServerSerializer 的另一种实现，采用 XML 格式将对象进行序列化/反序列化，见代码 7-13。

代码 7-13：类 ServerXMLSerializer

```
1  public class ServerXMLSerializer implements IServerSerializer<Request, Response>{
2      private XStream xStream;
3      public ServerXMLSerializer(){
4          xStream = new XStream();
```

```
5           xStream.alias(Request.class.getSimpleName(), Request.class);
6           xStream.alias(Response.class.getSimpleName(), Response.class);
7       }
8       @Override
9       public byte[] marshalling(Response response) throws Exception {
10          String xml = xStream.toXML(response);
11          return xml.getBytes("UTF-8");
12      }
13      @Override
14      public Request unmarshalling(byte[] data) throws Exception {
15          String xml = new String(data, "UTF-8");
16          return (Request) xStream.fromXML(xml);
17      }
18 }
```

7.3.2 反射层

反射层是根据客户端所请求的服务名称(类型为字符串),找到对应的服务实例并执行客户端所请求的方法(方法名称也是字符串)。代码7-14的RequestHandler类使用Java的反射机制完成对服务方法的调用。

代码7-14:类RequestHandler

```
1  @Data
2  @AllArgsConstructor
3  public class RequestHandler{
4      private ServiceRegister serviceRegister;
5      public Response handleRequest(Request req) throws Exception{
6          System.out.println("req = " + req);
7          //查找服务对象
8          ServiceObject so = this.serviceRegister.getServiceObject(
                                  req.getServiceName(), req.getVersion());
9          System.out.println(so);
10         Response rsp = null;
11         if (so == null) {
12             rsp = new Response(Status.NOT_FOUND);
13         } else {
14             //反射调用对应的过程方法
15             try {
16                 Method m = so.getInterf().getMethod(req.getMethod(),
                                          req.getPrameterTypes());
17                 Object returnValue = m.invoke(so.getObj(), req.getParameters());
18                 rsp = new Response(Status.SUCCESS);
19                 rsp.setReturnValue(returnValue);
```

```
20                } catch (NoSuchMethodException | SecurityException |
                          IllegalAccessException | IllegalArgumentException |
                          InvocationTargetException e) {
21                    rsp = new Response(Status.ERROR);
22                    rsp.setException(e);
23                }
24            }
25            return rsp;
26        }
27 }
```

代码 7-14 的第 8 行 getServiceObject(.)方法的定义见代码 7-15 的第 10～17 行。本例中，服务端所有的服务实例已经在启动时保存于 ConcurrentLinkedQueue 中，见代码 7-15 的第 2 行与第 7 行。

代码 7-15：类 ServiceRegister

```
1  public class ServiceRegister {
2      private ConcurrentLinkedQueue< ServiceObject> serviceDB =
                                           new ConcurrentLinkedQueue<>();
3      public void register(ServiceObject so, String protocol, int port) throws Exception{
4          if (so == null){
5              throw new IllegalArgumentException("参数不能为空!");
6          }
7          this.serviceDB.add(so);
8          //参数 protocol 与 port 在真正注册时使用,这里不进行注册
9      }
10     public ServiceObject getServiceObject(String interfName, String version){
11         for(ServiceObject so:serviceDB){
12             if (so.getInterf().getName().equals(interfName) &&
                                   so.getVerseion().equals(version)){
13                 return so;
14             }
15         }
16         return null;
17     }
18 }
```

服务端用 ServiceObject 来描述服务，其定义见代码 7-16。

代码 7-16：类 ServiceObject

```
1  @Data
2  @AllArgsConstructor
3  public class ServiceObject {
4      //interf 和 version 构成联合主键
5      private Class<?> interf;
```

```
6       private String verseion;
7       private Object obj;
8  }
```

第 7 行的 Object 对象是真正的服务实例。

7.3.3 通信层

RPC 框架服务端必须在系统启动时开启服务端口，等待 RPC 框架客户端的请求。定义抽象类 RpcServer，见代码 7-17。

代码 7-17：类 RpcServer

```
1  @Data
2  @AllArgsConstructor
3  public abstract class RpcServer {
4       protected int port;
5       protected String protocol;
6       protected RequestHandler handler;
7       protected IServerSerializer<Request, Response> serializer;
8       //启动服务
9       public abstract void start();
10      //停止服务
11      public abstract void stop();
12 }
```

抽象类 RpcServer 代表 RPC 框架服务端整体，所以第 6 行有 RequestHandler 作为成员变量，并且第 7 行有序列化器作为成员变量。该抽象类的子类通过重载 start()方法实现特定的通信层。代码 7-18 基于 Netty 实现了服务端通信层。

代码 7-18：类 NettyRpcServer

```
1  public class NettyRpcServer extends RpcServer {
2       public NettyRpcServer(int port, String protocol,
                             RequestHandler handler,
                             IServerSerializer<Request, Response> serializer) {
3           super(port, protocol, handler, serializer);
4       }
5       private Channel channel;
6       @Override
7       public void start() {
8           RecvHandler recvHandler = new RecvHandler(this.handler, this.serializer);
9           //配置服务器
10          EventLoopGroup bossGroup = new NioEventLoopGroup(1);
```

```
11          EventLoopGroup workerGroup = new NioEventLoopGroup();
12          try {
13              ServerBootstrap b = new ServerBootstrap();
14              b.group(bossGroup, workerGroup)
15                      .channel(NioServerSocketChannel.class)
16                      .localAddress(new InetSocketAddress(port))
17                      .option(ChannelOption.SO_BACKLOG, 100)
18                      .childHandler(new ChannelInitializer<SocketChannel>() {
19                  @Override
20                  public void initChannel(SocketChannel ch) throws Exception {
21                      ch.pipeline().addLast(new LengthFieldBasedFrameDecoder(
                                                   Integer.MAX_VALUE, 0, 4, 0, 4));
22                      ch.pipeline().addLast(recvHandler);
23                      ch.pipeline().addLast(new AttachHeaderHandler());
24                  }
25              });
26              //启动服务
27              ChannelFuture f = b.bind().sync();
28              System.out.println("完成服务端端口绑定与启动……");
29              this.channel = f.channel();
30              //等待服务通道关闭
31              f.channel().closeFuture().sync();
32          } catch (Exception e) {
33              e.printStackTrace();
34          } finally {
35              //释放线程组资源
36              bossGroup.shutdownGracefully();
37              workerGroup.shutdownGracefully();
38          }
39      }
40      @Override
41      public void stop() {
42          this.channel.close();
43      }
44      @ChannelHandler.Sharable
45      private class RecvHandler extends ChannelInboundHandlerAdapter {
46          private RequestHandler handler;
47          private IServerSerializer<Request, Response> serializer;
48          public RecvHandler(RequestHandler handler,
                          IServerSerializer<Request, Response> serializer){
49              this.handler = handler;
50              this.serializer = serializer;
51          }
```

```
52          @Override
53          public void channelRead(ChannelHandlerContext ctx, Object msg)
            throws Exception {
54              ByteBuf msgBuf = (ByteBuf) msg;
55              byte[] reqBytes = new byte[msgBuf.readableBytes()];
56              msgBuf.readBytes(reqBytes);
57              Request request = this.serializer.unmarshalling(reqBytes);
58              Response response = this.handler.handleRequest(request);
59              byte[] rspBytes = this.serializer.marshalling(response);
60              ByteBuf rspBuf = Unpooled.buffer(rspBytes.length);
61              rspBuf.writeBytes(rspBytes);
62              ctx.channel().write(rspBuf);
63          }
64          @Override
65          public void exceptionCaught(ChannelHandlerContext ctx, Throwable cause) {
66              cause.printStackTrace();
67              ctx.close();
68          }
69      }
70  }
```

第 57 行通过反序列化得到 request 对象，第 58 行调用反射器执行真正的服务函数并将函数值存放于 response 对象，第 59 行对 response 对象进行序列化，这三行代码是框架服务端的核心功能所在。

服务端通信层的类图见图 7-3。

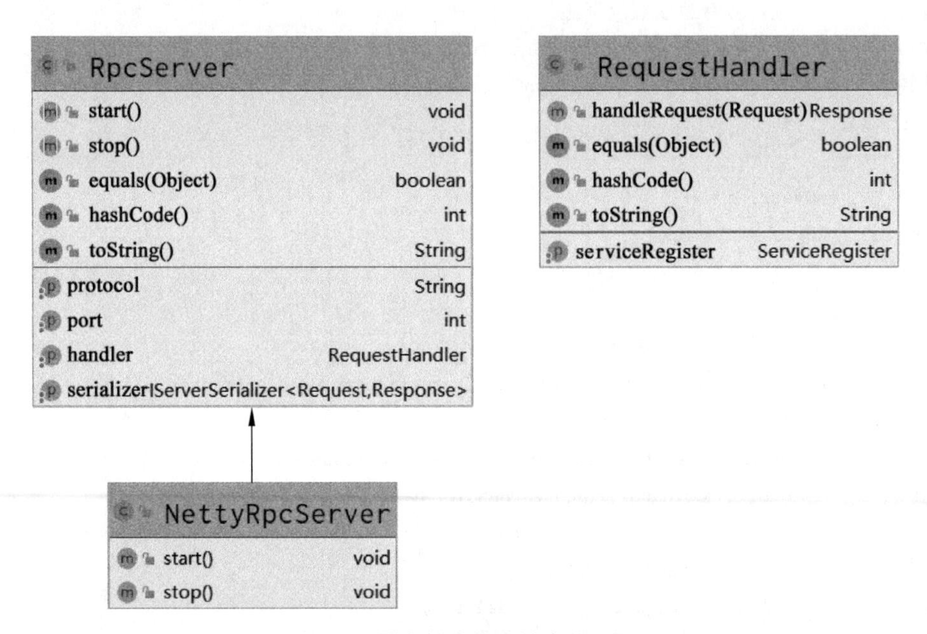

图 7-3　服务端通信层的类关系图

7.4 服务消费者

在客户端定义两个用于业务的服务接口，分别参见代码 7-19 和代码 7-20。这里只是接口，无须具体实现。这两个接口必须与服务发布者所发布的服务信息一致。

代码 7-19：接口 HelloInterface

```
1  public interface HelloInterface {
2      String sayHello(String name);
3      Point multiPoint(Point p, int multi);
4  }
```

代码 7-20：接口 StudentInterface

```
1  public interface StudentInterface {
2      List<Student> getAll();
3      Student getOne(Integer id);
4      boolean modifyStudent(Student student);
5      boolean eraseStudent(Integer id);
6      Student createStudent(Student student);
7  }
```

类 DefaultServiceInfoDiscovery 模拟服务发现，见代码 7-21。

代码 7-21：类 DefaultServiceInfoDiscovery

```
1  public class DefaultServiceInfoDiscovery implements ServiceInfoDiscoverer {
2      static String serializingType = PropertiesUtils.getProperties("rpc.serializing");
3      static String address = "127.0.0.1:" + PropertiesUtils.getProperties("rpc.port");
4      static List<ServiceInfo> serviceDB = Arrays.asList(
5              new ServiceInfo("cxiao.sh.cn.common.business.HelloInterface",
                          "1.0", address, serializingType),
6              new ServiceInfo("cxiao.sh.cn.common.business.HelloInterface",
                          "2.0", address, serializingType),
7              new ServiceInfo("cxiao.sh.cn.common.business.HelloInterface",
                          "3.0", address, serializingType),
8              new ServiceInfo("cxiao.sh.cn.common.business.StudentInterface",
                          "1.0", address, serializingType)
9      );
10     @Override
11     public List<ServiceInfo> getServiceInfo(String interfName, String version) {
12         List<ServiceInfo> list = new ArrayList<>();
13         for(ServiceInfo sInfo : serviceDB){
```

```
14            if (sInfo.getInterfName().equals(interfName) &&
                                                sInfo.getVersion().equals(version)){
15                list.add(sInfo);
16            }
17        }
18        return list;
19    }
20 }
```

框架客户端提供描述服务信息的类ServiceInfo，此类的定义见代码7-22。服务消费者在服务发现时获得该类的实例。

代码7-22：类ServiceInfo

```
1  @Data
2  @AllArgsConstructor
3  @NoArgsConstructor
4  public class ServiceInfo {
5      //因为服务信息来自注册中心，所以这里必须用字符串，不能用 Class <?>
6      //interfName 与 version 构成联合主键
7      private String interfName;
8      private String version;
9      private String address; //ip:port 的格式
10     private String serializingType;
11 }
```

服务消费者基于RPC框架的客户端进行服务调用，见代码7-23。

代码7-23：类Consumer

```
1  public class Consumer {
2      public static void main(String[] args) {
3          ClientStubProxyFactory cspf = new ClientStubProxyFactory();
4          //设置模拟的服务发现者，创建所支持的全部序列化器
5          cspf.setSid(new DefaultServiceInfoDiscovery());
6          Map<String, IClientSerializer<Request, Response>> supportingSerializers =
                                                         new HashMap<>();
7          supportingSerializers.put("javas", new ClientJavaSerializer());
8          supportingSerializers.put("xml", new ClientXMLSerializer());
9          cspf.setSupportingSerializers(supportingSerializers);
10         //设置网络层实现
11         cspf.setNetClient(new NettyNetClient());
12         //获取远程服务代理
13         HelloInterface hello = cspf.getProxy(HelloInterface.class, "2.0");
14         //像调用本地方法一样执行远程方法
15         String msg = hello.sayHello("World!");
16         System.out.println(msg);
```

```
17          System.out.println(hello.multiPoint(new Point(5,10), 2));
18          StudentInterface stUtils = cspf.getProxy(StudentInterface.class, "1.0");
19          System.out.println(stUtils.getAll());
20          System.out.println(stUtils.getOne(102));
21      }
22 }
```

第 13 行通过 getProxy(.)方法得到一个代理对象 hello，之后就如同使用本地对象一样调用该代理对象的方法。第 18 行类似，但是换了服务接口。

7.5 服务发布者

服务发布者使用 RPC 框架服务端，见代码 7-24。发布者必须提供服务的具体实现。

代码 7-24：类 Provider

```
1  public class Provider {
2      public static void main(String[] args) throws Exception {
3          //所有的对象用相同的 port
4          int port = Integer.parseInt(PropertiesUtils.getProperties("rpc.port"));
5          String serializingType = PropertiesUtils.getProperties("rpc.serializing");
6          //服务注册
7          ServiceRegister sr = new ServiceRegister();
8          List<ServiceObject> objects = Arrays.asList(
9                  new ServiceObject(HelloInterface.class, "1.0", new HelloService1()),
10                 new ServiceObject(HelloInterface.class, "2.0", new HelloService2()),
11                 new ServiceObject(HelloInterface.class, "3.0", new HelloService3()),
12                 new ServiceObject(StudentInterface.class,
                                           "1.0", new StudentService())
13         );
14         for (ServiceObject so:objects){
15             sr.register(so, serializingType, port);
16         }
17         //设置所支持的全部协议
18         Map<String, IServerSerializer<Request, Response>>
                                      supportingSerializers = new HashMap<>();
19         supportingSerializers.put("javas", new ServerJavaSerializer());
20         supportingSerializers.put("xml", new ServerXMLSerializer());
21         //获取所设置的序列化格式
22         IServerSerializer<Request, Response> serverSerializer =
                                   supportingSerializers.get(serializingType);
23         RequestHandler reqHandler = new RequestHandler(sr);
24         RpcServer server = new NettyRpcServer(port, serializingType,
                                           reqHandler, serverSerializer);
```

```
25          server.start();
26          System.in.read(); //按任意键退出
27          server.stop();
28      }
29 }
```

运行前先统一修改 RpcServer 项目与 RpcClient 项目的配置文件 app.properties，都改成如下形式。

```
rpc.port = 8001
rpc.serializing = javas
```

或者都改成如下形式。

```
rpc.port = 8001
rpc.serializing = xml
```

之后先启动 RpcServer 项目的 Provider 类，再启动 RpcClient 项目的 Consumer 类。

习题

为本章案例增加 JSON 序列化器。

第 8 章 两个简单应用

本章介绍两个简单应用：一个是基于 Spring Boot 的 WebSocket 应用，对应附件源码的 websocket 项目；另一个是邮件发送程序，对应附件源码的 SendMail 项目。

8.1 WebSocket 应用

WebSocket(RFC6455)消除了 HTTP 连接的无状态特性，提供了在一条 TCP 信道中客户端和服务端之间的全双工通信。使用 Spring Boot 开发 WebSocket 应用相对简单。这一节将用 WebSocket 做一个简易聊天室。

首先，添加依赖，如代码 8-1 所示。

代码 8-1：WebSocket 应用的依赖项

```
1  <dependencies>
2      <dependency>
3          <groupId>org.springframework.boot</groupId>
4          <artifactId>spring-boot-starter-web</artifactId>
5          <version>2.3.3.RELEASE</version>
6      </dependency>
7      <dependency>
8          <groupId>org.springframework.boot</groupId>
9          <artifactId>spring-boot-starter-websocket</artifactId>
10         <version>2.3.3.RELEASE</version>
11     </dependency>
12 </dependencies>
```

扩展 TextWebSocketHandler 类，重载其中若干 WebSocket 事件响应函数，见代码 8-2。

代码 8-2：类 EchoHandler

```
1  @Component
2  public class EchoHandler extends TextWebSocketHandler {
```

```
3       public static final ConcurrentLinkedQueue<WebSocketSession> clients =
                    new ConcurrentLinkedQueue<WebSocketSession>();
4       private static final DateTimeFormatter sf =
                    DateTimeFormatter.ofPattern("yyyy-MM-dd HH:mm:ss");
5       @Override
6       public void afterConnectionEstablished(WebSocketSession session)
        throws Exception {
7           //产生随机名称,只发送给当前连接者
8           String randName = getRandomString(8);
9           session.sendMessage(new TextMessage(randName + " 建立连接!"));
10          //保存名称,以便群发时使用
11          Map<String, Object> map = session.getAttributes();
12          map.put("name", randName);
13          //添加至连接者列表
14          clients.add(session);
15      }
16      @Override
17      protected void handleTextMessage(WebSocketSession session,
                                    TextMessage message) throws Exception {
18          //当服务端接收到任一连接者提交的消息时
19          var msg = message.getPayload();
20          //稍作加工,转发给全部的连接者。转发信息包含提交者的随机名称
21          for (WebSocketSession client:clients){
22              String strMsg = LocalDateTime.now().format(sf) ;
23              client.sendMessage(new TextMessage(strMsg));
24              strMsg = session.getAttributes().get("name") +  " : " + msg;
25              client.sendMessage(new TextMessage(strMsg));
26          }
27      }
28      @Override
29      public void afterConnectionClosed(WebSocketSession session, CloseStatus status)
        throws Exception {
30          super.afterConnectionClosed(session, status);
31          clients.remove(session);
32      }
33      //生成指定 length 的随机字符串(A~Z,a~z,0~9)
34      public static String getRandomString(int length) {
35          String str = "abcdefghijklmnopqrstuvwxyzABCDEFGHIJKLMNOPQRSTUVWXYZ0123456789";
36          Random random = new Random();
37          StringBuffer sb = new StringBuffer();
38          for (int i = 0; i < length; i++) {
39              int number = random.nextInt(str.length());
40              sb.append(str.charAt(number));
```

```
41            }
42            return sb.toString();
43        }
44 }
```

当服务端发生与WebSocket通信相关的事件时，这个类中相应的方法将被自动调用。每当一个新的客户端与服务端建立WebSocket连接，服务端就会得到一个WebSocketSession对象，这个WebSocketSession对象就是第6行函数的参数。第12行给新建立的连接设置一个随机字符串的名称，第14行把新建立的连接加入一个ConcurrentLinkedQueue对象。

当服务端接收到任一WebSocket连接者发送的消息时，第17行的handleTextMessage(.)方法将被自动调用，该方法的第一个参数表示发送消息的连接者，第二个参数是所发送的消息。第21～26行把接收到的任何一条消息都转发给已经记录在案的所有连接者，转发的消息还包括当前时间和原始发送者的名称，这个名称是在连接建立之初服务端给连接者设置的随机字符串。

当服务端的某个WebSocket连接被关闭的时候，第29行的afterConnectionClosed(.)方法将被自动调用，该方法的第一个参数表示所关闭的连接，第二个参数是关闭状态码，第31行把已关闭的连接从ConcurrentLinkedQueue对象里移除，以确保ConcurrentLinkedQueue对象里保存的都是可用连接。

服务端启动类见代码8-3。

代码8-3：类MyApplication

```
1  @SpringBootApplication
2  @EnableWebSocket
3  public class MyApplication implements WebSocketConfigurer {
4      //启动后，在浏览器输入 http://localhost:8080
5      public static void main(String[ ] args) {
6          SpringApplication.run(MyApplication.class, args);
7      }
8      @Autowired
9      EchoHandler echoHandler;
10     @Override
11     public void registerWebSocketHandlers(
                           WebSocketHandlerRegistry webSocketHandlerRegistry) {
12         webSocketHandlerRegistry.addHandler(echoHandler, "/echo");
13     }
14 }
```

在启动类头部标注@EnableWebSocket，第8行注入EchoHandler实例，第11行重载方法registerWebSocketHandlers(.)指定EchoHandler实例对应的URI，当客户端访问包

含此URI的地址时，将与服务端建立具有持久性的WebSocket连接。

客户端使用的静态网页包括两个：代码8-4的app.js和代码8-5的index.html，都放在resources/static/文件夹内。

代码8-4：app.js

```
1  var ws = null;
2  var url = "ws://localhost:8080/echo";
3  function setConnected(connected) {
4      document.getElementById('connect').disabled = connected;
5      document.getElementById('disconnect').disabled = !connected;
6      document.getElementById('echo').disabled = !connected;
7  }
8  function connect() {
9      ws = new WebSocket(url);
10     ws.onopen = function () {
11         setConnected(true);
12     };
13     ws.onmessage = function (event) {
14         log(event.data);
15     };
16     ws.onclose = function (event) {
17         setConnected(false);
18         log('提示：关闭连接。');
19     };
20 }
21 function disconnect() {
22     if (ws != null) {
23         ws.close();
24         ws = null;
25     }
26     setConnected(false);
27 }
28 function echo() {
29     if (ws != null) {
30         var message = document.getElementById('message').value;
31         log('发送：' + message);
32         ws.send(message);
33     } else {
34         alert('尚未建立连接，请单击"连接"按钮建立连接！');
35     }
36 }
37 function log(message) {
38     var console = document.getElementById('logging');
39     var p = document.createElement('p');
```

```
40      p.appendChild(document.createTextNode(message));
41      console.appendChild(p);
42      while (console.childNodes.length > 12) {
43          console.removeChild(console.firstChild);
44      }
45      console.scrollTop = console.scrollHeight;
46 }
```

代码 8-5：index.html

```
1  <!DOCTYPE html>
2  <html>
3  <head>
4  <meta http-equiv="Content-Type" content="text/html; charset=UTF-8" />
5  <link type="text/css" rel="stylesheet"
     href="https://cdnjs.cloudflare.com/ajax/libs/semantic-ui/2.2.10/semantic.min.css" />
6  <script type="text/javascript" src="app.js"></script>
7  </head>
8  <body>
9  <div>
10 <div id="connect-container" class="ui centered grid">
11 <div class="row">
12 <button id="connect" onclick="connect();" class="ui green button ">连接</button>
13 <button id="disconnect" disabled="disabled" onclick="disconnect();" class=
     "ui red button">断开连接</button>
14 </div>
15 <div class="row">
16 <textarea id="message" style="width: 350px" class="ui input" placeholder=
     "要发送的消息"></textarea>
17 </div>
18 <div class="row">
19 <button id="echo" onclick="echo();" disabled="disabled" class="ui button">
     发送消息</button>
20 </div>
21 </div>
22 <div id="console-container">
23 <h3>Logging</h3>
24 <div id="logging"></div>
25 </div>
26 </div>
27 </body>
28 </html>
```

运行时启动源码中项目 websocket 的 MyApplication 类，打开两个浏览器，都访问网址 http://localhost:8080/，可以模拟两人对话，如图 8-1 所示。

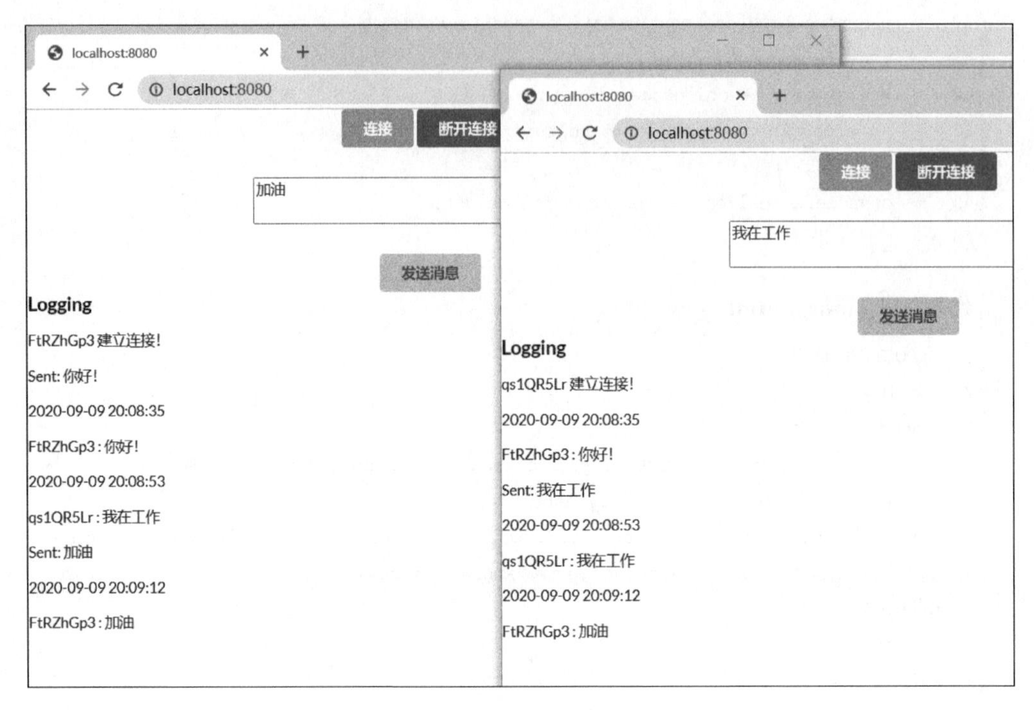

图 8-1　WebSocket 应用效果图

8.2　邮件发送程序

本节将用程序来发送电子邮件。

首先导入依赖项，见代码 8-6。

代码 8-6：邮件发送程序的依赖项

```
1  <dependencies>
2      <dependency>
3          <groupId>com.sun.mail</groupId>
4          <artifactId>javax.mail</artifactId>
5          <version>1.6.2</version>
6      </dependency>
7      <dependency>
8          <groupId>org.projectlombok</groupId>
9          <artifactId>lombok</artifactId>
10         <version>1.18.10</version>
11     </dependency>
12 </dependencies>
```

定义 SimpleMail 类，用于存放邮件基本信息，见代码 8-7。

代码 8-7：类 SimpleMail

```
1  @Data
2  public class SimpleMail {
3      private String[] to;
4      private String[] cc;
5      private String[] bcc;
6      private String from;
7      private String host;
8      private String port;
9      private String subject;
10     private String body;
11     private String[] attachment;
12     private String username;
13     private String password;
14     private Boolean isHtml;
15 }
```

定义 MailUtils 类，里面的 sendMail(.)函数用于发送邮件，见代码 8-8。

代码 8-8：类 MailUtils

```
1  public class MailUtils {
2    public static void sendMail(SimpleMail simpleMail) {
3      Properties properties = System.getProperties();
4      properties.setProperty("mail.smtp.host", simpleMail.getHost());
5      properties.setProperty("mail.smtp.port", simpleMail.getPort());
6      properties.put("mail.smtp.auth", "true");
7      try {
8          Session session = Session.getDefaultInstance(properties, new Authenticator() {
9              @Override
10             protected PasswordAuthentication getPasswordAuthentication() {
11                 return new PasswordAuthentication(
                               simpleMail.getUsername(), simpleMail.getPassword());
12             }
13         });
14         MimeMessage message = new MimeMessage(session);
15         message.setFrom(new InternetAddress(simpleMail.getFrom()));
16         List<InternetAddress> toList = new ArrayList<>();
17         List<InternetAddress> ccList = new ArrayList<>();
18         List<InternetAddress> bccList = new ArrayList<>();
19         if (simpleMail.getTo() != null) {
20             for (String str : simpleMail.getTo()) {
21                 toList.add(new InternetAddress(str));
22             }
23             message.addRecipients(Message.RecipientType.TO,
                                   toList.toArray(new InternetAddress[]{ }));
```

```
24          }
25          if (simpleMail.getCc() != null) {
26              for (String str : simpleMail.getCc()) {
27                  ccList.add(new InternetAddress(str));
28              }
29              message.addRecipients(Message.RecipientType.CC,
                                  ccList.toArray(new InternetAddress[]{ }));
30          }
31          if (simpleMail.getBcc() != null) {
32              for (String str : simpleMail.getBcc()) {
33                  bccList.add(new InternetAddress(str));
34              }
35              message.addRecipients(Message.RecipientType.BCC,
                                  bccList.toArray(new InternetAddress[]{ }));
36          }
37          message.setSubject(simpleMail.getSubject());
38          BodyPart messageBodyPart = new MimeBodyPart();
39          Multipart multipart = new MimeMultipart();
40          if (simpleMail.getIsHtml()) {
41              messageBodyPart.setContent(simpleMail.getBody(),
                                      "text/html;charset = UTF-8");
42          } else {
43              messageBodyPart.setContent(simpleMail.getBody(),
                                      "text/plain;charset = UTF-8");
44          }
45          multipart.addBodyPart(messageBodyPart);
46          if (simpleMail.getAttachment() != null) {
47              for (String str : simpleMail.getAttachment()) {
48                  messageBodyPart = new MimeBodyPart();
49                  DataSource source = new FileDataSource(str);
50                  messageBodyPart.setDataHandler(new DataHandler(source));
51                  messageBodyPart.setFileName(MimeUtility.encodeText(
                                      str.substring(str.lastIndexOf("\\") + 1)));
52                  multipart.addBodyPart(messageBodyPart);
53              }
54          }
55          message.setContent(multipart);
56          Transport.send(message);
57          System.out.println("发送成功!!!!!");
58      }catch (MessagingException | UnsupportedEncodingException mex) {
59          mex.printStackTrace();
60          System.out.println("发送失败：" + simpleMail.getTo()[0]);
61      }
62    }
63 }
```

启动类见代码 8-9。

代码 8-9：类 MailClient

```
1  public class MailClient {
2      public static void main(String[ ] args) {
3          SimpleMail simpleMail = new SimpleMail();
4          simpleMail.setHost("smtp.qq.com");
5          simpleMail.setPort("25");
6          simpleMail.setUsername("184088321");
7          simpleMail.setPassword("buzzrzrfjdhmcahd");
8          simpleMail.setFrom("184088321@qq.com");
9          simpleMail.setTo(new String[ ]{"123456@qq.com","cxiao@fudan.edu.cn"});
10         simpleMail.setCc(null);
11         simpleMail.setBcc(null);
12         simpleMail.setSubject("Test Mail");
13         simpleMail.setBody("Hello! I am xiaochuan from qq.");
14         simpleMail.setAttachment(new String[ ]{"D:\\ClientFiles\\abc.txt",
                                                 "D:\\ServerFiles\\xyz.pdf"});
15         simpleMail.setIsHtml(false);
16         MailUtils.sendMail(simpleMail);
17     }
18 }
```

先设置代码 8-9 中 simpleMail 的各项，再运行。

习题

为本章的 WebSocket 案例增加用户认证功能。

附　　录

1. 将本地jar包导入本地仓库并交给maven管理

将jar包放入任意文件夹,cmd执行如下命令。

```
mvn install:install-file
    -Dfile=commons-logging-1.1.1.jar
    -DgroupId=pay
    -DartifactId=commons-logging-1.1.1
    -Dversion=1.0.0
    -Dpackaging=jar
```

命令解释如下。

Dfile：是要转成的maven依赖的jar包。

DgroupId：决定了将要添加至本地仓库的位置。

DartifactId：可以随便起一个名称(用来标识maven包的ID)。

Dversion：将要生成的版本(并不是本地jar包的版本)设置成1.0.0。因为将要生成pay的maven包并不存在,这将是第一个。

G、A、V这3项参数在本地仓库可以随意指定,但若要公用则还是采用jar自带的GAV。

任何一个jar包在用maven打包时必然需要指定G(groupId)、A(artifactId)、V(version),上述这条命令的参数中就使用了GAV。

命令运行后,界面出现"BUILD SUCCESS",表示导入成功。

2. 生产环境使用线程池示例

```
ExecutorService fixPool = new ThreadPoolExecutor(
    10, 20, 500L, TimeUnit.MILLISECONDS,new LinkedBlockingQueue<Runnable>(20));
```

3. 解决jar包运行时显示"没有主清单属性"问题

在pom.xml文件中添加如下插件,注意execution条目不能少,然后重新打包。

```
<build>
    <plugins>
        <plugin>
```

```
                <groupId>org.springframework.boot</groupId>
                <artifactId>spring-boot-maven-plugin</artifactId>
                <version>2.2.6.RELEASE</version>
                <executions>
                    <execution>
                        <goals>
                            <goal>repackage</goal>
                        </goals>
                    </execution>
                </executions>
            </plugin>
        </plugins>
</build>
```

4. 在 Intellij IDEA 复制已有的模块(Module)

按如下 4 个步骤操作。

第 1 步：复制模块文件夹到项目文件夹，可以通过两种途径实现：一种途径是在操作系统上复制；另一种途径是在 IDEA 内右键单击要复制的模块，选择“Copy”，再右键单击“External Libraries”，选择“Paste”，这样就完成了模块文件夹的复制。模块文件夹的名称将是模块名，但是此时并没有把新模块加入项目。

第 2 步：进入操作系统，修改新模块文件夹内的 *.iml 文件名，改成新的模块名。

第 3 步：选择菜单 File→New→Module from Existing Sources，选择第 2 步改名的 .iml 文件，再单击按钮“OK”。这里后缀为.iml 的文件是 Intellij IDEA module file。如果第 3 步选择的是 pom.xml 文件，则引导窗口还有若干设置步骤，而选择 .iml 文件则不需要另外设置。

第 4 步：导入模块后，修改 pom.xml 中的<artifactId>节点，使之与模块名相同。这一步很重要。